Essence of Eternity II:
Quaternion Dimension – Fermion Duality in SuperStandard Theories

Stephen Blaha Ph. D.
Blaha Research

One-to-One Fermion-Dimension Duality Established
256 QUeST Fermions
Four Species of QUeST Fermions
Any Fermion is Transformable to Any Other Fermion Using Symmetry Groups
Particle Functionals Support Relativistic Instantaneous Entanglement
Particle Functionals, Monads, Observability
Observability and Absolute Reality

Pingree-Hill Publishing

MMXX

ISBN: 978-1-7345834-6-5

Rev. 00/00/01 July 4, 2020

To Margaret

Some Other Books by Stephen Blaha

All the Megaverse! Starships Exploring the Endless Universes of the Cosmos using the Baryonic Force (Blaha Research, Auburn, NH, 2014)

SuperCivilizations: Civilizations as Superorganisms (McMann-Fisher Publishing, Auburn, NH, 2010)

All the Universe! Faster Than Light Tachyon Quark Starships & Particle Accelerators with the LHC as a Prototype Starship Drive Scientific Edition (Pingree-Hill Publishing, Auburn, NH, 2011).

Unification of God Theory and Unified SuperStandard Model THIRD EDITION (Pingree Hill Publishing, Auburn, NH, 2018).

The Exact QED Calculation of the Fine Structure Constant Implies ALL 4D Universes have the Same Physics/Life Prospects (Pingree Hill Publishing, Auburn, NH, 2019).

Unified SuperStandard Theory and the SuperUniverse Model: The Foundation of Science (Pingree Hill Publishing, Auburn, NH, 2018).

Quaternion Unified SuperStandard Theory (The QUeST) and Megaverse Octonion SuperStandard Theory (MOST) (Pingree Hill Publishing, Auburn, NH, 2020).

Unified SuperStandard Theories for Quaternion Universes & The Octonion Megaverse (Pingree Hill Publishing, Auburn, NH, 2020).

The Essence of Eternity: Quaternion & Octonion SuperStandard Theories (Pingree Hill Publishing, Auburn, NH, 2020).

Available on Amazon.com, bn.com Amazon.co.uk and other international web sites as well as at better bookstores (through Ingram Distributors).

CONTENTS

FIGURES and TABLES

INTRODUCTION

In previous books this author has derived the Unified SuperStandard Theory (UST) in our 3 + 1 dimension space-time from Complex General Relativity and Quantum Field Theory suitably extended. Recently we showed that 32 complex quaternion dimension QUeST gives the identical pattern of Internal Symmetries as UST. *UST is derivable from QUeST.*

This remarkable coincidence leads us to explore Unified SuperStandard Theories in greater detail.

In this book we examine particle-dimension duality and find that the one-to-one match extends down to the individual fermion and dimension level.

We then develop the monad – particle functional concept that had enabled us to eliminate the issues of instantaneous quantum entanglement with Special Relativity in earlier books.

We also defined a complex octonion space of 32 dimensions called MOST in previous books. It also has a close (fermion) particle-dimension duality.

These considerations led us to consider the implications of monads. We found it provides a basis for entanglement and observability. It leads to an absolute reality throughout the universe using particle functionals to give observability at every point in the universe. At each point of space-time gravitons, fermions and bosons provide obervability.

The Megaverse also has a close fermion-dimension duality. We find the Megaverse is also has total observability and absolute reality.

1. Dimensions of Unified SuperStandard Theories

Dimensions are usually thought to be static—existing only to be used to define the coordinates of physical theories. They are not thought to have a dynamic aspect. In this chapter we will define the dimensions of the Unified SuperStandard Theory (UST) in our space-time, and in an underlying complex quaternion space that we call QUeST. We will see that the internal symmetries of UST emerge directly from the set of dimensions of QUeST. Thus internal symmetry dimensions of UST and the Standard Model are no longer a subject of mystery and no longer unusual (as they are sometimes portrayed.)

The fundamental UST theory is described in detail in Blaha (2020c) (and earlier books. Its basis in 32 dimension complex quaternion space (QUeST) is described in Blaha (2020d) – also in detail. It is remarkable that QUeST provides an exact fundamental basis for UST.[1]

1.1 QUeST Dimensions

QUeST is defined in a 32 dimension complex quaternion space. There are a total of 256 individual dimensions in this space. In view of the number of dimensions the usual approach of defining coordinates is cumbersome. Consequently we followed a Mathematical Picture Language approach in Blaha (2020d). This approach was originally used by Pythagoras and His School around 500 BCE.[2] Pythagoras used • symbols, which he called *psiphi* symbols (meaning pebbles).

We will begin with an 8 dimension complex quaternion space which has 64 dimensions. Then we will consider the 32 dimension complex quaternion space which can be viewed as consisting of four layers of 8 dimension complex quaternion space.

---

[1] The basis of UST in QUeST was not known to the author until Fall, 2019 although the form of UST was known to the author many years earlier and recorded in several books.

[2] Kirk (1962) presents much of what is known of the Pythagoreans. This author developed the psiphi diagrams, originally, without being aware of the Pythagorean diagrammatic language.

We express the 8-dimension space as a diagram in Fig. 1.1. It consists of a pattern of psiphi. Then we will partition it into the dimensions of space-time and internal symmetry groups in the next section.

•••• ••••
•••• ••••
•••• ••••
•••• ••••
•••• ••••
•••• ••••
•••• ••••
•••• ••••

Figure 1.1. Psiphi diagram of the dimensions of 8 dimension complex quaternion space. Each row represents a complex quaternion with 8 dimensions.

Similarly the 32 dimension complex quaternion space is depicted as in Fig. 1.2.

•••• ••••
•••• ••••
•••• ••••
•••• ••••
•••• ••••
•••• ••••
•••• ••••
•••
•••• ••••

Figure 1.2. Psiphi diagram of the dimensions of 32 dimension complex quaternion space. Each row again represents a complex quaternion with 8 dimensions.

Fig. 1.1 is a psiphi diagram that represents 7 + 1 coordinates with 8 dimensions:

Time Biquaternion

$$t = (a + ib + jc + kd) + I(a' + ib' + jc' + kd') \tag{1.1}$$

Spatial Biquaternions

$$\begin{aligned}
x &= (a_x + ib_x + jc_x + kd_x) + I(a'_x + ib_x' + jc_x' + kd_x') \\
y &= (a_y + ib_y + jc_y + kd_y) + I(a'_y + ib_y' + jc_y' + kd_y') \\
z &= (a_z + ib_z + jc_z + kd_z) + I(a'_z + ib_z' + jc_z' + kd_z') \\
x1 &= (a_{x1} + ib_{x1} + jc_{x1} + kd_{x1}) + I(a'_{x1} + ib_{x1}' + jc_{x1}' + kd_{x1}') \\
y1 &= (a_{y1} + ib_{y1} + jc_{y1} + kd_{y1}) + I(a'_{y1} + ib_{y1}' + jc_{y1}' + kd_{y1}') \\
z1 &= (a_{z1} + ib_{z1} + jc_{z1} + kd_{z1}) + I(a'_{z1} + ib_{z1}' + jc_{z1}' + kd_{z1}') \\
w1 &= (a_{w1} + ib_{w1} + jc_{w1} + kd_{w1}) + I(a'_{w1} + ib_{w1}' + jc_{w1}' + kd_{w1}')
\end{aligned}$$

where all coefficients: a, b, c, d, a′, b′, c′, d′, and a_i, b_i, c_i, d_i, a'_i, b'_i, c'_i, d'_i for i = x, y, z, w, x1, y1, z1, w1 are *real-valued* numbers, and where I is an additional fundamental quaternion unit that makes each quaternion "complex." Note that the real and imaginary part of each coordinate has the same fundamental quaternion units to permit complex rotations between them.

As we will see we need four iterations of the above set of complex quaternions for the space of QUeST so that it will yield UST upon restriction to real-valued coordinates. Thus QUeST requires a 32 dimension complex quaternion space:

$$\begin{aligned}
t &= (a + ib + jc + kd) + I(a' + ib' + jc' + kd') \qquad (1.2) \\
x &= (a_x + ib_x + jc_x + kd_x) + I(a'_x + ib_x' + jc_x' + kd_x') \\
y &= (a_y + ib_y + jc_y + kd_y) + I(a'_y + ib_y' + jc_y' + kd_y') \\
z &= (a_z + ib_z + jc_z + kd_z) + I(a'_z + ib_z' + jc_z' + kd_z') \\
x1 &= (a_{x1} + ib_{x1} + jc_{x1} + kd_{x1}) + I(a'_{x1} + ib_{x1}' + jc_{x1}' + kd_{x1}') \\
y1 &= (a_{y1} + ib_{y1} + jc_{y1} + kd_{y1}) + I(a'_{y1} + ib_{y1}' + jc_{y1}' + kd_{y1}') \\
z1 &= (a_{z1} + ib_{z1} + jc_{z1} + kd_{z1}) + I(a'_{z1} + ib_{z1}' + jc_{z1}' + kd_{z1}') \\
w1 &= (a_{w1} + ib_{w1} + jc_{w1} + kd_{w1}) + I(a'_{w1} + ib_{w1}' + jc_{w1}' + kd_{w1}') \\
&\ldots \\
w4 &= (a_{w4} + ib_{w4} + jc_{w4} + kd_{w4}) + I(a'_{w4} + ib_{w4}' + jc_{w4}' + kd_{w4}')
\end{aligned}$$

1.2 Breakup of QUeST Space into Space-time and Internal Symmetry Groups

Returning now to psiphi diagrams we find that the 8-dimension complex quaternion space can be partitioned into blocks based on the dimension rules:

U(2) requires 4 dimensions
U(1)⊗SU(2) requires 4 dimensions
SU(3) requires 6 dimensions
U(4) requires 8 dimensions

where the dimensions have real-valued coordinates and are called *real dimensions*. Fig. 1.3 shows the partition of one layer QUeST (8 dimensions by eq. 1.1) giving eq. 1.3.

$$SU(2)\otimes U(1)\otimes SU(3)\otimes U(2)\otimes SU(2)\otimes U(1)\otimes SU(3)\otimes U(2) \tag{1.3}$$

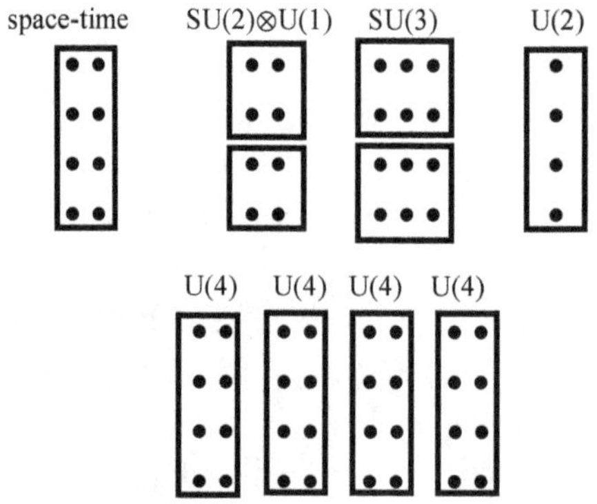

Figure 1.3. Psiphi diagram showing partitioning of 8 dimension complex quaternion space. The blocks of dimensions yield 4 complex dimension space-time, and the internal symmetries of normal matter and Dark matter SU(2)⊗U(1)⊗SU(3)⊗U(2)⊗SU(2)⊗U(1)⊗SU(3)⊗U(2) as they appear in UST. The lower U(4) groups are for the Generation group and the Layer group for normal matter and for Dark matter in one layer UST. The U(2) group transforms between normal and Dark matter.

1.3 Justification for a Four Layer QUeST

There is good reason for QUeST to have four layers embodied in 32 dimension complex quaternion space. If one considers the content of the layer displayed in Fig. 1.3 one sees a 4 dimension complex coordinates block for space-time. To create a 4 dimension complex quaternion coordinates space-time, one needs four layers of the form of Fig. 1.3. *The combination of the four 4-dimension complex coordinate parts is a complex quaternion dimensions space-time*. Thus the choice of four layer QUeST gives us a 4-dimension complex quaternion space-time AND enables QUeST to map directly to UST with its four layers if one limits the quaternion coordinates to the real-valued coordinates within them.[3] If one did not define a four layer QUeST then the role of the four U(4) Layer Groups would be in doubt since the Layer Groups in UST transform among each of the four generations of fermions. (See Fig. 1.6 for an illustration.) Four generations implies a need for four Layer groups, which the 32 complex quaternion QUeST contains. Note four U(2) groups that transform between normal and Dark sectors for each layer are required. We call these groups *Dark* groups.

We conclude four layer QUeST is needed to have a 4 dimension complex quaternion space-time.

1.4 One Layer QUeST Structure

As a preliminary to four layer 32 complex quaternion dimensions QUeST we display the structure implicit in Fig. 1.3 in Fig. 1.4.

---

[3] The Layer groups of UST enable mixing between the layers of fermions as shown in Fig. 2.1.

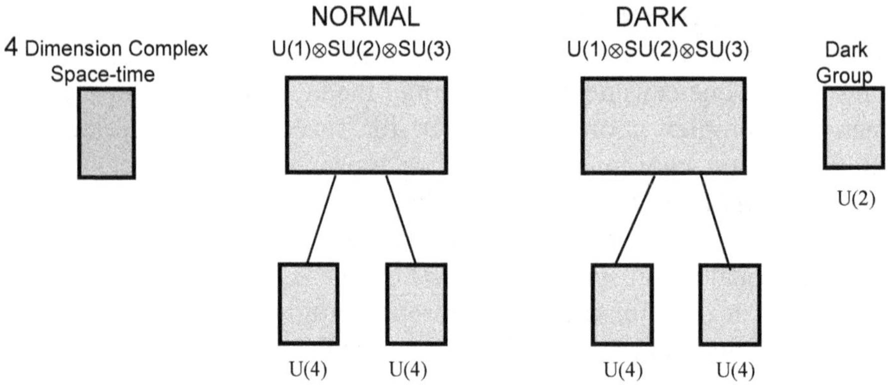

Figure 1.4. Schematic of the structure of the internal symmetry groups of eq. 1.3 plus 4 complex dimensions space-time. The two large blocks are each 5 dimension complex coordinate representations of SU(2)⊗U(1)⊗SU(3). The U(2) group supports transformations (rotations) between normal and Dark matter.

1.5 Four Layer QUeST

Fig. 1.5 shows four layer QUeST internal symmetry groups and 4 dimension complex quaternion space-time. Fig. 1.7 shows the 4 layer fundamental fermion spectrum. Fig. 3.1 shows the map between 32 dimension complex quaternion dimensions and fundamental fermions. The map is one-to-one for all four layers.

The 256 dimensions of the 32 dimension complex quaternion space equal the 256 fundamental fermions of QUeST and UST.

The internal symmetry group structure of Fig. 1.5 is

$$[SU(2)\otimes U(1)\otimes SU(3)\otimes SU(2)\otimes U(1)\otimes SU(3)\otimes U(4)^4\otimes U(2)]^4 \qquad (1.4)$$

plus 4 dimension complex quaternion space-time.

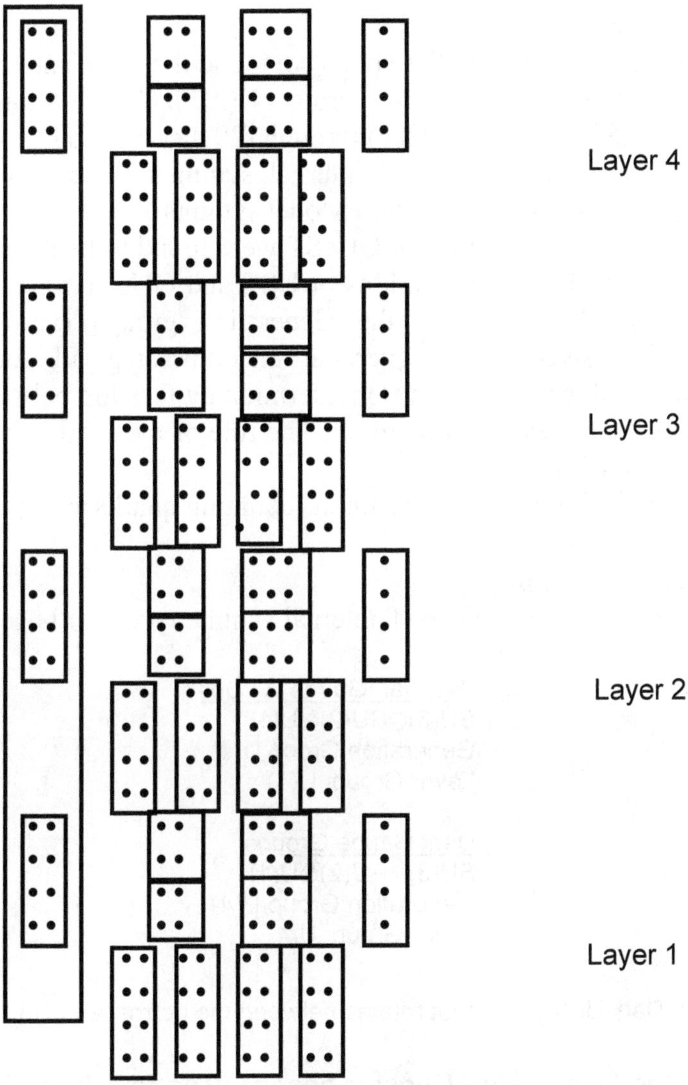

Figure 1.5. Four layer QUeST internal symmetry groups and space-time diagram for 32 dimension complex quaternion space. Note the left composite blocks combine to specify a 4 dimension complex quaternion space-time.

1.6 Particles of UST and QUeST

In Blaha (2020c) and earlier books we found the set of internal symmetry groups of eq. 1.4 to which we added the Dark groups $U(2)^4$ based on a fundamental derivation of UST from QUeST in Blaha (2020a) through (2020c). QUeST provides a fundamental basis for UST and Standard Model internal symmetries removing the questions of strangeness often attributed to Standard Model symmetries.

The fundamental fermions of QUeST were found to be the same as in UST. Fig. 1.7 displays the four layers of fermions of UST and QUeST together with the roles of the SU(2)⊗U(1)⊗SU(3) groups, the Generation groups (vertical within fermion generations), the Layer groups (vertical encompassing all four layers for each generation), the Dark groups (one-to-one, fermion by fermion between normal and Dark sectors), and the Complex Lorentz group. The role of these groups will be discussed in more detail in chapter 8.

There are 256 fundamental fermions counting quarks as triplets.

1.7 QUeST Vector Bosons

The overall *one layer* QUeST internal symmetry vector bosons are:

<u>"Normal" Gauge Groups</u>
SU(3)⊗SU(2)⊗U(1)
Generation Group U(4)
Layer Group U(4)

<u>Dark Gauge Groups</u>
SU(3)⊗SU(2)⊗U(1)
Generation Group U(4)
Layer Group U(4)

PLUS

A Dark U(2) group that rotates between the normal and Dark sectors

Figure 1.6. One layer QUeST vector bosons. The four layer QUeST quadruples the above list: with one distinct set for each layer.

The Fermion Periodic Table

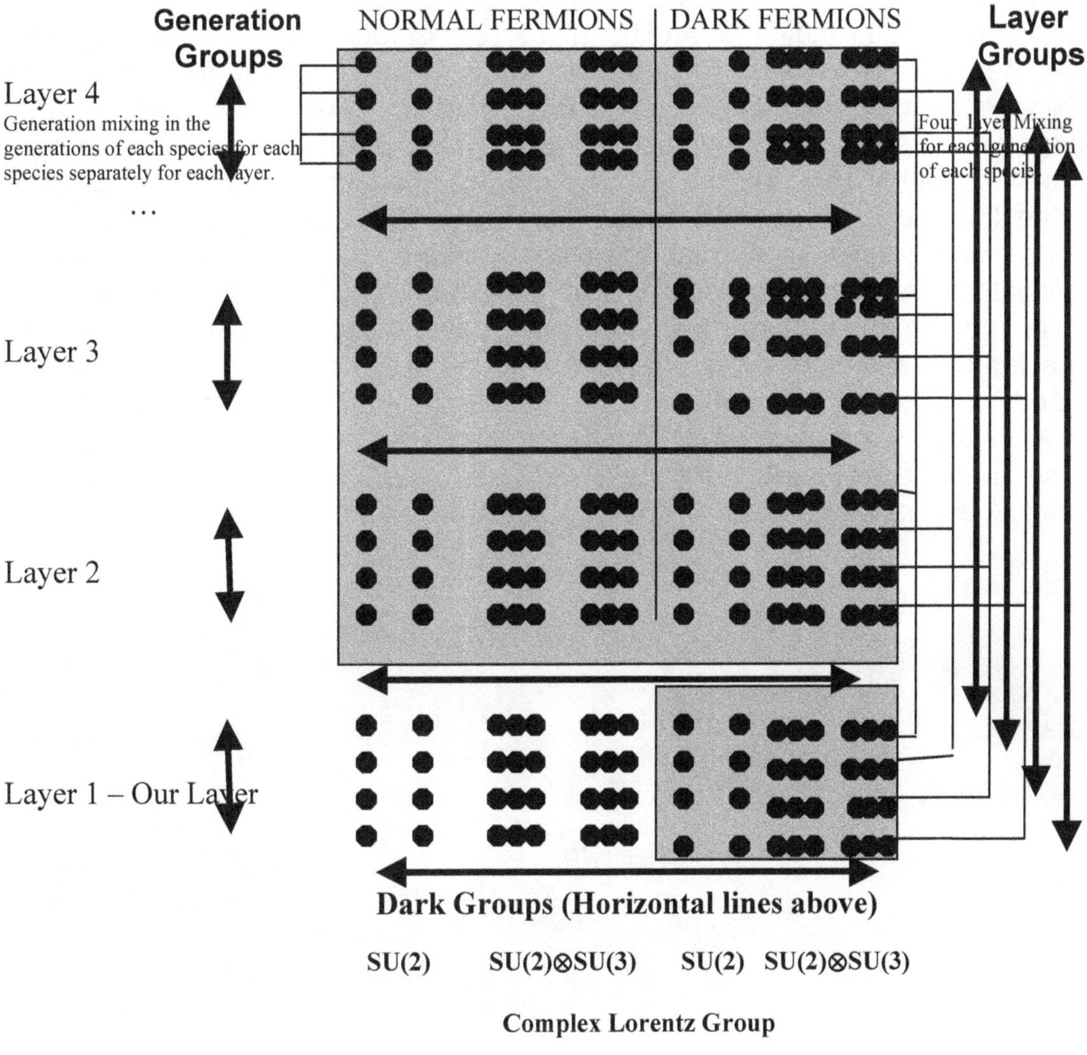

Figure 1.7. Fermion particle spectrum and partial example of pattern of mass mixing of the Generation, Layer, and Dark grroups. Unshaded parts are the known fermions including an additional, as yet not found, 4th generation shown. The lines on the left side (only shown for one layer) display the Generation mixing within each layer's species. The Generation mixing applies within each layer using a separate Generation group for each layer. The lines on the right side show Layer group mixing with the mixing amongst all four layers for each of the four generations individually. There are four Layer groups. The Dark groups mixing between normal and Dark fermions are shown in the center as horizontal lines. For each generation and each layer SU(2) mixes between an e-type fermion and a neutrino-type fermion. It also mixes between an up-quark-type fermion and a down-quark-type fermion. SU(3) mixes among each up-quark triplet and down-quark triplet separately. Complex Lorentz group transformations map among all four fermions: Dirac ↔ tachyon ↔ up-quark ↔ down-quark. See Fig. 8.1 for details. There are 256 fundamental fermions counting quarks as triplets.

2. Megaverse 32 Complex Octonion Space (MOST)

The 32 complex octonion space of the Megaverse (or Multiverse) is factored into a 7 complex quaternion space-time and a set of internal symmetries.[4]The Magaverse can contain our universe as well as a host of other universes.

We find it convenient to split the Megaverse into four 8 complex octonion subspaces. These subspaces are duplicates of each other but contain different internal symmetry groups and different MOST fermion and vector boson spectrums.

Fig 2.1 symbolically depicts an 8-dimension complex octonion (bioctonion) space with a psiphi • for each real-valued dimension. *We treat bioctonion space as a higher dimensional space and do not use details of octonion algebra in our development.*

Figure 2.1. Eight-Dimensional (7 + 1) complex octonion subspace with coordinates represented by • 's. This subspace has 128 real dimensions.

The internal symmetry dimensions above number 112. Sixteen of the dimensions serve as the dimensions of an 8-dimension complex space-time. These

[4] Much of this chapter appears in Blaha (2020d).

dimensions serve as the fundamental representation dimensions[5] of each of the factors of

$$[SU(2)\otimes U(1)\otimes SU(3)\otimes SU(2)\otimes U(1)\otimes SU(3)]^2\otimes U(4)^9 \quad (2.1)$$

counting a U(4) Dark group.

Thc U(4) Generation and Layer groups are represented in Fig. 2.1. We depict the pattern of symmetry implied by Fig. 2.1 and eq. 2.1 in Fig. 2.2 and 2.3 below.

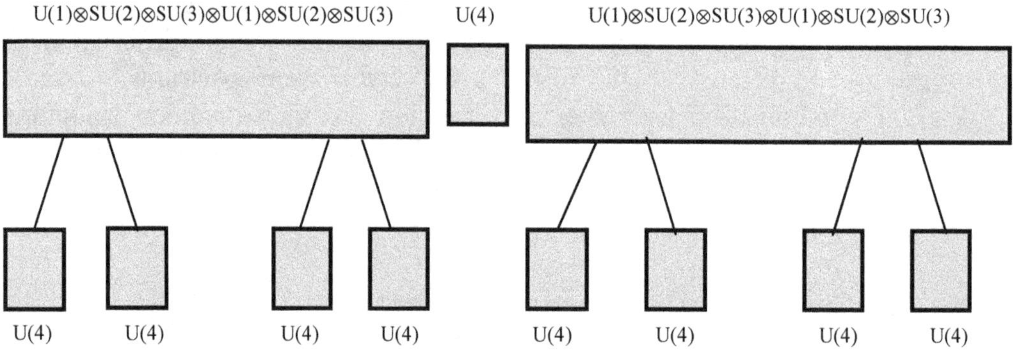

Figure 2.2. Schematic of the internal symmetry groups' dimensions of Fig. 2.1. The two "large" blocks are each sets of 20 real[6] dimensions furnishing representations of the indicated groups. The lower U(4) groups are the Generation and Layer number groups. The Dark U(4) group is shown. The total number of real dimensions is 112.

Each U(1)⊗SU(2)⊗SU(3)⊗U(1)⊗SU(2)⊗SU(3) block in Fig. 2.3 has a 10 complex dimensions (20 real dimensions) representation. The blocks are subdivided in Fig. 2.3 into sets of 10 real dimensions supporting representations of U(1)⊗SU(2)⊗SU(3). We assign the first block to contain the representations of the known parts of the Standard Model. There are three Dark blocks. The internal symmetry groups of each part are listed in Fig. 2.3.

[5] See section 1.2.
[6] We also use the term complex dimensions to indicate pairs of real dimensions.

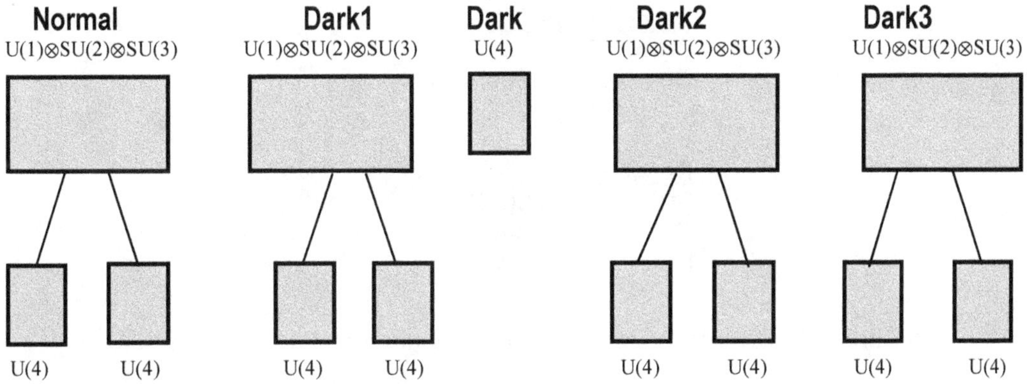

Figure 2.3. Schematic of the internal symmetry groups of eq. 2.1 including the Dark U(4) group. These are the internal symmetry groups of one layer MOST. The lower U(4) groups above are the Generation and Layer number groups. One pair of each number group is for each of the four U(1)⊗SU(2)⊗SU(3) factors above.

2.1 One Layer MOST

The above section specifies the *one layer* MOST. The symmetries of the three other layers are the same but their groups, and fermions, are individual to each layer. The groups of each layer can be flagged with a different index.

The overall one layer MOST internal symmetry is specified by Fig. 2.3, eq. 2.1, and an 8-dimension complex space-time. The internal symmetry groups of one Layer MOST are:

<u>"Normal" Gauge Groups</u>
SU(3)⊗SU(2)⊗U(1)
Generation Group U(4)
Layer Group U(4)
<u>Dark1 Gauge Groups</u>
SU(3)⊗SU(2)⊗U(1)
Generation Group U(4)
Layer Group U(4)

Dark2 Gauge Groups
$SU(3) \otimes SU(2) \otimes U(1)$
Generation Group U(4)
Layer Group U(4)

Dark3 Gauge Groups
$SU(3) \otimes SU(2) \otimes U(1)$
Generation Group U(4)
Layer Group U(4)

PLUS

A Dark U(4) group that rotates among the four normal and Dark sectors

Figure 2.4. One layer MOST vector bosons list from eq. 2.1. The four layer MOST quadruples the above list: with one distinct set for each layer. In one layer the total number of vector bosons of the above list is 192. Thus four layers yield a total count of 768 vector bosons in MOST (not counting the Species group which comes from General Relativity). We require each layer has a separate Dark U(4) rotation group.

2.2 Four Layer MOST

The four layer MOST is described by a 32 dimension complex octonion space. Thus it consists of four "copies" of the coordinates:

Figure 2.5. The 32 dimension MOST schematic. Four layer MOST has 512 real dimensions.

They yield four duplicates of the internal symmetry schematic in Fig. 2.3, and an 8 complex quaternion space-time consisting of 7 + 1 complex-valued quaternion coordinates (obtained by combining the four layers of 8 dimension complex space-time coordinates.)

The sum total of real dimensions is 512 as is the sum of the dimensions of the above parts constructed from the dimensions.

2.3 Justification for a Four Layer MOST

There is good reason for MOST to have four layers embodied in 32 dimension complex quaternion space. If one considers the content of a layer one sees a 8-dimension complex coordinates block for space-time. To create an 8 dimension complex *quaternion* coordinates space-time, one needs four layers. *The combination of four 8-dimension complex coordinates is an 8 dimension complex quaternion space-time.*

Thus the choice of four layer MOST gives us an 8-dimension complex quaternion space-time that can contain 4-dimension complex quaternion QUeST universes. *We conclude four layer MOST is needed to have an 8-dimension complex quaternion space-time.*

The thirty-two dimension complex octonion space contains an 8-dimension complex quaternion space-time and the four layers of Internal Symmetry groups shown in Fig. 7.5.

2.4 Fermion and Gauge Vector Boson Spectrums

The fermion and vector boson spectrums that emerge in MOST are those of an "enlarged" QUeST and Unified SuperStandard Theory. They are displayed below. MOST has an additional two Dark sectors beyond QUeST and the Unified SuperStandard Theory.

Vector Bosons

From Fig. 2.4 we find MOST has 192 vector bosons in one layer. Thus four layer MOST has a total count of 768 MOST vector bosons. There are two additional Dark vector boson sectors beyond QUeST and the Unified SuperStandard Theory.

Fermions

There are 512 fundamental fermions in MOST, which includes two additional Dark fermion sectors. Fig. 2.6 shows the MOST fermion spectrum.

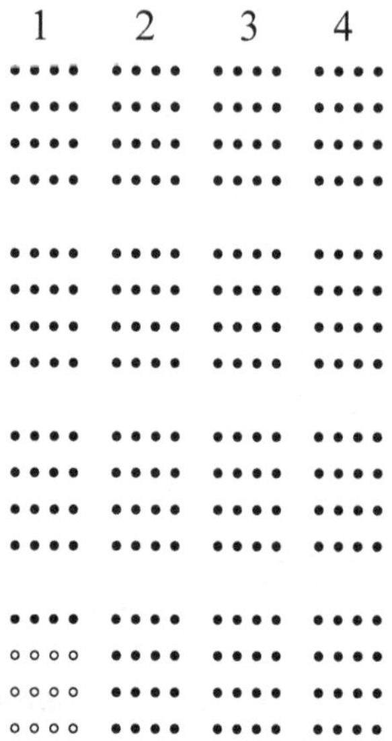

Figure 2.6. Schematic spectrum of the fermions of 4 layer MOST. Each fermion is represented by a •. Quark triplets are represented by a single •. Four sets of four species in four generations which are in turn in 4 layers. Open symbols ∘ represent known fermions. There are 512 fundamental fermions taking account of quark triplets. Note the Layer groups determine the layers in UST. They require 4 layers of 8 complex octonions in Megaverse space leading to the 32 dimension complex octonion space.

3. Particle-Dimension Duality

3.1 Particle-Dimension Duality in our QUeST Universe

There is a remarkable correspondence between the dimensions of 32 dimension complex quaternion space and the fermion spectrum of UST and QUeST. Fig. 3.1 shows the map between the dimensions and fermions for one layer (8 complex quaternion dimensions and one layer of fermions in UST.) The other three layers of dimensions and fermions exhibit the same map. Consequently there is a one-to-one correspondence between the 256 dimensions and the 256 fundamental fermions in QUeST.

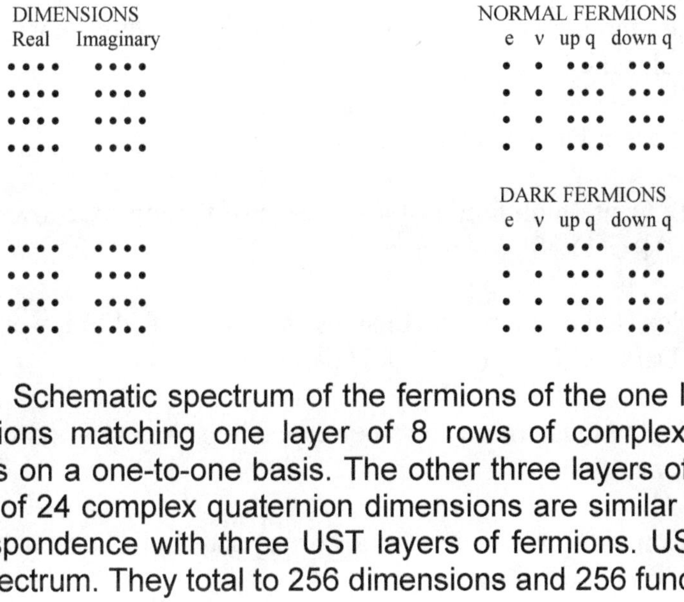

Figure 3.1. Schematic spectrum of the fermions of the one layer of normal and Dark fermions matching one layer of 8 rows of complex quaternion space dimensions on a one-to-one basis. The other three layers of space dimensions consisting of 24 complex quaternion dimensions are similar in having a one-to-one correspondence with three UST layers of fermions. UST has a four layer fermion spectrum. They total to 256 dimensions and 256 fundamental fermions.

The map is one-to-one and applies to individual particles and dimensions. Labeling the top row of dimensions of Fig. 3.1 as d_1, d_2, …, d_8 we see the duality in detail for the top row of Fig. 3.1:

$$
\begin{aligned}
e &\leftrightarrow d_1 \\
\nu &\leftrightarrow d_2 \\
q_1 &\leftrightarrow d_1 d_4 \\
q_2 &\leftrightarrow d_1 d_5 \\
q_3 &\leftrightarrow d_1 d_6 \\
q_4 &\leftrightarrow d_2 d_4 \\
q_5 &\leftrightarrow d_2 d_5 \\
q_6 &\leftrightarrow d_2 d_6
\end{aligned} \tag{3.1}
$$

The quarks have both SU(2)⊗U(1) and SU(3) symmetry resulting in the products of dimensions in the correspondence. (We assume free particles in these discussions.)

The one-to-one map applies to the rows of Fig. 3.1 for both normal and Dark sectors. Fig. 3.1 illustrates the following points:

1. A One-To-One Map: Dimensions – Fermions.

2. The 8 Complex Quaternion Dimensions in each of the top four rows correspond to normal SU(2)⊗U(1)⊗SU(3) particles.

3. The 8 Complex Quaternion Dimensions in each of the lower four rows correspond to Dark SU(2)⊗U(1)⊗SU(3) particles.

4. The 4 top Complex Quaternion rows corresponding to normal particles implies four fermion generations.

5. The 4 lower Complex Quaternion rows corresponding to Dark particles imply four Dark fermion generations.

6. 4 Layers of 8 Complex Quaternion rows imply 4 layers of fermions as seen in UST.

Thus our QUeST formalism accounts for the number of fermions per generation, the number of generations per layer (4) for normal and Dark fermions, the number of normal and Dark layers (4), the number of space-time dimensions (4), and the number of complex quaternion dimensions[7] (32).

3.2 Particle-Dimension Duality in our MOST Megaverse

There is also a one-to-one correspondence between the dimensions of 32 dimension complex octonion space and the fermion spectrum of MOST. Fig. 3.2 shows the map between the dimensions and fermions for one layer (8 complex octonion dimensions and one layer of fermions in MOST.) The other three layers of dimensions and fermions exhibit the same map. As a result there is a one-to-one correspondence between 512 dimensions and 512 fundamental fermions in MOST – the Unified SuperStandard Theory of the Megaverse.

3.3 Implications of Particle-Dimension Duality

Particle-dimension duality combined with the definition of particle fields in terms of functionals (See Blaha (2020c) and earlier books.), and a map of dimensions to coordinate dimensions of fundamental group representations, and thence to functionals enables us to define a triality of dimensions, functionals and particles. The triality may be symbolized by

$$d_i \leftrightarrow g_i \leftrightarrow f_i$$

where d_i is a dimension, g_i is the dimension taken to be a dimension of a group's fundamental representation, and f_i is a functional which is subject to the group's transformations.

We consider the implications of these considerations in succeeding chapters.

---

[7] Eq. 1.2.

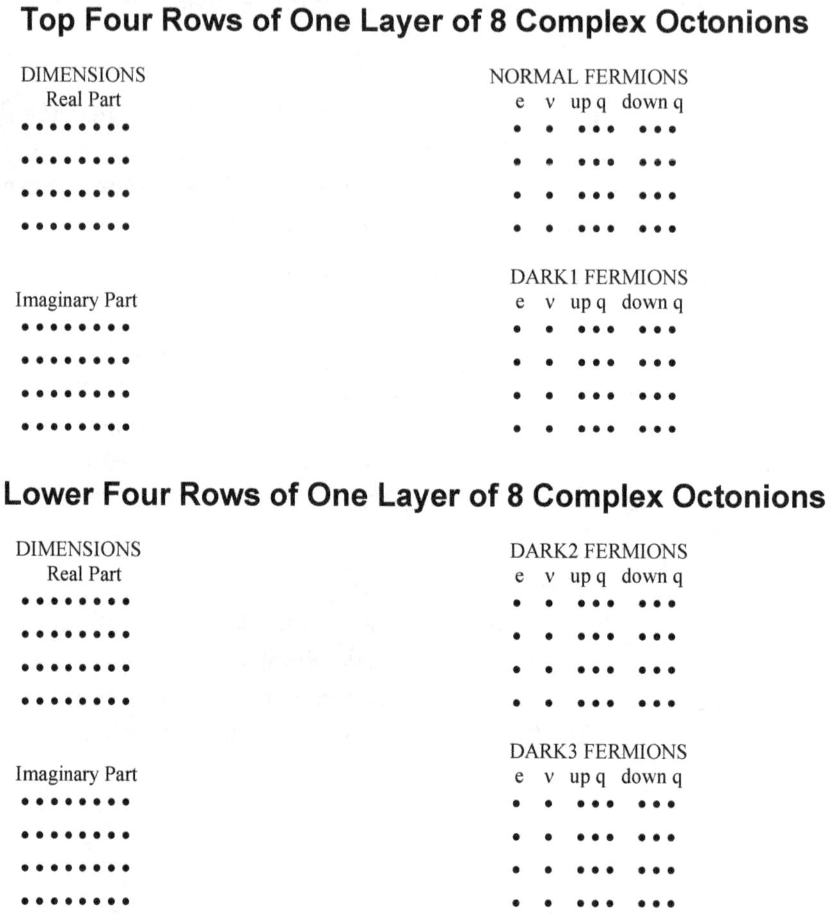

Figure 3.2. Schematic spectrum of the fermions of the one layer of normal and Dark fermions matching one layer of 8 rows of complex octonion space (See Fig. 2.1) on a one-to-one basis. The other three layers of space dimensions consisting of 24 complex octonion dimensions are similar in having a one-to-one correspondence with three UST layers of fermions. UST has a four layer fermion spectrum. They total to 512 dimensions and 512 fundamental fermions.

4. Status of Elementary Particle Theory

Elementary particle theory has gone through a number of phases in the past forty years. Forty years ago model quantum field theories largely based on quantum field theory and various forms of internal symmetries. The motivation for these theories was the success of the Standard Model. It was thought more was needed.

The resulting theories incorporated the Standard Model within generalizations of Standard Model internal symmetries. The primary issue was the justification for proposed internal symmetries.

There was also a deeper approach based on SuperString theory. SuperString theories encountered numerous issues. A primary issue was the selection of a SuperString theory that led inexorably to a physically realistic theory along the lines of the Standard Model. The search for the "true" SuperString theory, if it exists, has occupied the efforts of a sizeable number of theorists for forty years.

In view of the quagmire of theoretic efforts for a Theory of Elementary Particles this author developed a theoretic approach based on a set of fundamental postulates in the manner of Euclid. With six postulates the author was able to develop the known parts in the form of the Standard Model. Part of this development was the introduction of a modified quantum field theory formalism using Two Tier field theory (to eliminate divergences in perturbation theory) and PseudoQuantum field theory which supports higher derivative lagrangians in a canonical manner and enables reasonable physics in curved space-times. It also appeared reasonable to base the theory on Complex Special Relativity. Complex General Relativity became necessary in order to create a unified theory for curved space-time. The result was the Unified SuperStandard Theory (UST).

4.1 Unified SuperStandard Theory (UST)

UST had a fundamental fermion spectrum that contained the known fermion spectrum, and added a fourth generation, added four layers of fermions, and added a Dark matter spectrum of fermions that mirrored the normal fermion spectrum. The vector boson interactions contained the ElectroWeak interactions, U(4) interactions that caused interactions between fermions in each layer and similarly for Dark matter. The Generation group interactions yielded $U(4)^8$. The theory also had interactions between the layers of fermions.

UST also contained the Higgs boson sector and gravitational interactions according to a modified General Relativity. General Relativity and the Strong interactions were modified to have higher order derivative field equations. Consequently the gravitational force was modified in a manner similar to MoND. Standard Model interactions were modified to have a linear potential.

The theory is described in Blaha (2018f) and (2020c).

4.2 Quaternion Unified SuperStandard Theory (QUeST)

QUeST was defined in 32 dimension complex quaternion space. The study of QUeST[8] showed that it had the internal symmetries of UST plus an additional Dark U(2) interaction that transformed between normal particles and their Dark equivalents for all four fermion layers giving an additional $U(2)^4$ symmetry.

The space-time generated from QUeST a 4-dimension complex quaternion space that became real 4-dimension space-time in UST when quaternions were restricted to real coordinates.

Thus we saw that UST is properly viewed as derivative from QUeST. This unanticipated result solidified our feeling that QUeST and UST are the true theories of elementary particles.

4.3 Potential QUeST Perturbation Theory Divergences

QUeST is defined in a 4-dimension complex quaternion space-time. Just as 4-dimension complex space-time is restricted to real-valued 4-dimension space-time, QUeST is restricted to the 4-dimension real-valued space-time of UST.

The question then arises of high energy infinities in quaternion space-time since it has 32 real-valued dimensions. Integrations in momentum space in this space-time have the general form

$$\int d^{32}k$$

---

[8] See Blaha (2020a) through (2020d)/

Since particle propagators tend to be inverse quadratic (or inverse cubic) it appears that perturbation theory integrations will be highly ultraviolet divergent.

However if one uses Two-Tier coordinates as we do in UST an exponentially convergent factor in all dimensions appears eliminating ultra-violet divergences. In Two-Tier quaternion theory coordinates are replaced with Two-Tier coordinates

$$X^{\mu} - x^{\mu} + iY^{\mu} /M^2$$

for i = 1, 2, …, 32 where M is a very large mass of the order of the Planck mass presumably. Thus the Two-Tier version of QUeST has convergent perturbation theory integrals.

4.4 Consequences

The smooth transition from QUeST to UST increases the likelihood that QUeST is the true theory of elementary particles and gravitation. It encompasses all we know experimentally. And it has the ability to grow as more is learned about Higgs symmetry breaking.

If QUeST is the correct theory then elementary particle physics become similar to Chemistry. However there is still much to learn about detailed features and much room for the expansion of the understanding of the basis of the theory.

4.5 MOST Extension to the Megaverse/Multiverse

A remarkable aspect of the theoretical approach presented in Blaha (2020c), and here, is its ability to be extended to a Megaverse/Multiverse of universes. MOST showed that a 32 dimension complex octonion space leads to a 8 dimension complex quaternion space-time and an enlarged set of internal symmetries. Thus our universe, and unlimited numbers of additional universes, can "fit" into the MOST Megaverse. The internal symmetries of QUeST are a subset of the internal symmetries of MOST — a salutary feature.. Thus we have a fairly complete total picture of the universe and of the Megaverse.

4.6 Future Directions

Perhaps the most important issue facing particle theory is to obtain an understanding of the rationale for basing QUeST on complex quaternions. We know that Streater (2000) was able to justify extending space-time to complex-valued coordinates. What justification is there for complex quaternion coordinates? The number of these dimensions is 32. A partial justification for 32 dimensions is the need to obtain a 4-dimension complex quaternion space-time from which our real-valued space-time emerges.

Other tasks that remain are:

1. Determine the symmetry breaking for the internal symmetries including the symmetry breakdown(s) of 32 dimension complex quaternion space to internal symmetries and space-time.

2. Find the missing fermions and their interactions.

3. Determine the full set of symmetry breakdown parameters.

4. Find the deeper meaning of Higgs symmetry breaking from vacuum dynamics.

5. Develop a Chemistry of elementary particles.

5. Particle Functionals

5.1 Quantum Entanglement and Action-at-a-Distance

In 1935 Einstein, Podolsky, and Rosen[9] (EPR) raised questions about Quantum Entanglement. EPR considered the quantum entanglement of two systems and showed that instantaneous action-at-a-distance (spookiness) apparently resulted.

In this chapter we show how instantaneity is resolved by "factoring" wave functions into the composition of a functional and a Fourier wave function expansion. in In section 5.3 we show the instantaneity of the EPR two system state example is eliminated using functional factoring of wave functions.

Then we establish a functional formulation of the internal symmetry dimensions of 32 complex quaternion space. The functional formulation then generates fundamental representations of internal symmetry groups.

5.2 Functional Factorization of Quantum Fields

Many years ago Dirac factored the Klein-Gordan equation and obtained the Dirac equation for spin ½ fermions.

In the following sections we show that there is good reason to factor quantum mechanical wave functions and second quantized fields into an inner product of a particle functional and a corresponding Fourier coordinate expansion. With this factorization, and the assumption that the space of all particle functionals, as well as the space of all Fourier coordinate expansions, are both point spaces[10] with the consequence that there is no distance measure in either space, we find a change in one of a pair of space-like separated parts of an initial state causes an instantaneous transformation of the other part (eliminating Einstein's spookiness).

---

[9] Einstein A, Podolsky B, and Rosen N, "Can Quantum-Mechanical Description of Physical Reality Be Considered Complete?", Phys. Rev. **47**, 777 (1935).

[10] The points can be viewed as spaces with no distance measure that are factors in a tensor product with space-time.

5.3 Motivation for Quantum Field Factorization: Instantaneous Effects in Quantum Phenomena

Seemingly instantaneous quantum phenomena are apparent in many cases. For example:

1. Two particles placed in a definite spin state may separate to a space-like distance. If the z component of spin is flipped in one of the particles, the other particle instantaneously flips its spin in such a way as to conserve spin. This type of phenomena has been described as 'spooky' since it violates the law that no effect can travel at a rate faster than the speed of light.

2. Transitions between atomic levels take place instantaneously—in a zero time interval.

Quantum field factorization enables instantaneous effects to happen without violating Relativity.

5.4 General Form of Factorization

Normally fermion and boson quantum fields are described by a wave function of the form

$$\chi(\mathbf{x}, t) \tag{5.1}$$

We can formally factorize quantum fields as an inner product of a functional f_k and a space-time Fourier expansion denoted (k, **x**, t) (neglecting internal quantum numbers temporarily) where k is the momentum.

$$\chi(\mathbf{x}, t) = (f_k, (k, \mathbf{x}, t)) \tag{5.2}$$

For a free *two* particle wave function (non-interacting) the wave function may be written as a product of inner products:

$$\chi(\mathbf{x}, t) = (f_{1k}, (k, \mathbf{x}, t)_1)\, (f_{2q}, (q, \mathbf{x}, t)_2) \tag{5.3}$$

where k and q are momenta.

5.5 Factorization of Fermion Quantum Fields

We now consider the example of a free fermion field to illustrate the general concept. We began by defining a coordinate space Dirac Fourier quantum expansion as

$$(s, x, t) = N(p)[b(p, s)u(p, s)e^{-ip\cdot x} + d^\dagger(p, s)v(p, s)e^{+ip\cdot x}] \qquad (5.4)$$

where N(p) is a normalization factor, u and v are functions of spin and momentum, and b and $d^\dagger$ are creation/annihilation operators. We defined a Dirac quantum wave function with the inner product of a functional and a coordinate space Fourier quantum expansion:

$$\psi(x) = (f, (s, x, t)) = \sum_{\pm s} \int d^3pN(p)[b(p, s)u(p, s)e^{-ip\cdot x} + d^\dagger(p, s)v(p, s)e^{+ip\cdot x}] \qquad (5.5)$$

where we use a functional inner product formalism in the manner of Riesz (1955)[11] and others.

5.6 The EPR Two System State Example

EPR considered a state consisting of two systems that might become separated spatially. We can represent the state as

$$\Psi = \Sigma_n\psi_{1n}(x_1)\psi_{2n}(x_2) \qquad (5.6)$$

We can represent a measurement (reduction of state) with a projection Π_{1a} of system "1" to a state ψ_{1a} with

$$\psi_{1a} = \delta_{ab}\, \Pi_a\, \psi_{1b} \qquad (5.7)$$

Then

$$\Psi_{projected} = \Pi_{1a}\, \Sigma_n\psi_{1n}\psi_{2n} = \psi_{1a}(x_1)\psi_{2a}(x_2) \qquad (5.8)$$

---

[11] For example see pp. 61-2 of Riesz (1955) where linear functionals and their inner products are defined.

The effect of the measurement of system "1" is *instantaneous* of system "2" because the quantum functionals f_{1n} and f_{2n}, and the projections Π_{1n} and Π_{2n} of both systems are not separated by distance with the result

$$\psi_{1n}(x) = f_{1xn}(\Pi_{1xn}\Phi\,) \;= (f_{1xn}, \Pi_{1xn}\Phi) \qquad (5.9)$$
$$\psi_{2n}(y) = f_{2ny}(\Pi_{2yn}\Phi\,) \;= (f_{2yn}\,, \Pi_{2yn}\Phi) \qquad (5.10)$$

with $x = x_1$ and $y = x_2$. *The quantum functional and the projection select the wave and its coordinate parameterization. The coordinates in the wave are merely place holders.*

Therefore the relative distance between the coordinates x_1 and x_2 is not relevant for the change of state of system "2". The quantum functionals and projections give the instantaneity of the change in ψ_{2a} upon the measurement of system "1".

The EPR Spookiness is resolved by quantum functionals. There is no conflict with the Theory of Special Relativity.

5.7 Factorization Details

The rationale for factorization lies in the nature of the functionals and coordinate Fourier expansions that we use. For, we choose to create a space of particle functionals for fermions, bosons, and other particle states that consists of a single point with no distance measure (or alternately put, zero distance between all functionals.) We also choose to create a 'point' space of all coordinate Fourier expansions for bosons and fermions, whose elements have all coordinate values, x.

For the moment we wish to note that the space of functionals includes functionals for all fundamental particles, and all matter/energy composites, in the universe (and the Megaverse). We can describe transitions (interactions) in which functionals are "transformed" into other functionals. So the space of functionals has a dynamic aspect. Another important aspect of functional space is its universality—*all functionals of the Megaverse are present creating a type of link between all parts of the Cosmos.*

The space of coordinate Fourier expansions consists of all possible expansions for particles in the coordinates of each respective universe and of the Megaverse. This space also has no distance measure.

The factorization that we propose enables instantaneous communication of a transition between two space-like separated parts of a state. A change in one part immediately causes a corresponding change in the other part because the changes take place in the functionals which are located at the same point in functional space.

In a certain sense we have divorced quantum phenomena from coordinate space by quantum field factorization.

6. Functionals in QUeST's 32 Dimension Complex Quaternion Space

The dimensions of QUeST space can be initially treated simply as independent dimensions. Dimensions can be mapped to particles and thence to their particle functionals. These functionals can then be used to furnish fundamental representations of internal symmetry groups. They can also be used to define quantum fields for elementary particles as in eq. 3.1. Since some fundamental fermions, namely quarks, have transformation properties of both SU(3) and SU(2)⊗U(1) the quark functionals are equivalent to composites of dimensions. Thus eq. 3.1 becomes

Particle Functional		Dimension Composite	
e	$\leftrightarrow$	d_1	(6.1)
ν	$\leftrightarrow$	d_2	
q_1	$\leftrightarrow$	d_1d_4	
q_2	$\leftrightarrow$	d_1d_5	
q_3	$\leftrightarrow$	d_1d_6	
q_4	$\leftrightarrow$	d_2d_4	
q_5	$\leftrightarrow$	d_2d_5	
q_6	$\leftrightarrow$	d_2d_6	

for the SU(3)⊗SU(2)⊗U(1) set of fermions in each generation and each layer of normal and Dark fermions.

6.1 Form of a Fundamental Fermion Particle Functional

The general form of a fermion functional is

$$F_{fermion} = f_{internal}f_{spin} \tag{6.2}$$

where $f_{internal}$ is labeled with internal symmetry quantum numbers and f_{spin} is labeled by the spin state. They are factored to avoid SU6)-like problems found in the 1960s.

We will call functionals of the form of $F_{fermion}$ *fermion particle functionals.* We will call functionals of the form of $f_{Internal}$ *internal symmetry functionals* since they embody internal symmetries. We will call functionals of the form of f_{spin} *spin functionals* since they embody spin. In chapter 7 we will consider boson functionals.

6.2 Particle Functional Byte Numbering

The 256 fundamental fermions may be numbered from 1 through 256. The ASCII character tables have an 8-bit (byte) numbering. We may then represent a particle functional as

$$f_{byte} = f_{internal} \tag{6.3}$$

using bytes with values from 1 through 256.

Consequently fermions can be characterized as "letters" in an alphabet, and aggregates of fermions can be viewed as words. As a result one may think of the universe as a great Word—a concept which has been put forward many times.[12]

[12] A study of this possibility appears in Blaha (1998).

7. Four-Dimension Complex Quaternion Space-Time

QUeST is a theoretical foundation for the author's Unified SuperStandard Theory (UST). It is based on a 32 complex quaternion space with a total of 256 dimensions. Within this space there are a 4 dimension complex quaternion space-time with 32 dimensions, and a 224 dimension Internal Symmetry sector containing groups of UST including the U(2) Dark group..

The 4 dimension complex quaternion space-time is broken to a 4 complex dimension space-time, which becomes our space-time upon restriction to real dimensions (coordinates).

4 Dimension Complex Quaternion → 4 Dimension Complex → 4 Dimension Real

Figure 7.1. Progression of Space-times from QUeST to UST.

8. Complex Lorentz Group Boosts and Four Fermion Species

8.1 Complex Lorentz Boosts to Obtain the Four Species of Fermions

In UST[13] we showed that there are four Complex Lorentz group boosts that transform a spinor of a fermion at rest to four possible states: a Dirac electron-like fermion, a tachyon neutrino-like fermion, an up-quark-like fermion with real energy and complex 3-momentum, and a down-quark-like tachyon fermion with real energy and complex 3-momentum. We identified these fermions with the known leptons and quarks. We called each type of fermion a *species*. We showed that there was evidence to support faster-than-light neutrinos, and no evidence to prove down-type quarks were not tachyons. (In fact known fits to deep inelastic electron-nucleon data involve cutoffs that might be due to tachyon down quarks.)

In this chapter we examine the 4 dimension complex quaternion space-time of QUeST[14] and the complex space-time of UST (before restriction to real coordinates.) Each yields the four species of fermions.

The UST derivation of the four fermion species appears in Appendix A.

8.2 The Four QUeST Fermion Species

QUeST is defined in a 4-dimension complex quaternion space. We have extracted four complex-valued (eight real-valued) coordinates to form a space-time in chapter 7. Complex coordinates support a Complex Lorentz group just as the United SuperStandard theory does to some degree.

Appendix A describes the origin and features of the four species of fundamental fermions in UST in detail: "charged" lepton species, neutral lepton species, up-type quark species, and down-type quark species.

---

[13] Blaha (2020c) and earlier books such as Blaha (2007b).

[14] We also discuss the four species of Megaverse MOST.

The features of the QUeST fermion species are the same as in UST. The key relation in complex quaternion space is

$$e^x = e^a \,(\cos(||\mathbf{v}||) + \mathbf{v}/||\mathbf{v}|| \sin(||\mathbf{v}||))s \tag{8.1}$$

It is analogous to the similar complex coordinates identity used in Appendix A in eqs. 3.2, 3.3, 3.11, 3.12 and so on With it we can determine the boosts in 4 dimension complex quaternion space-time

8.3 QUeST Fermion Species

The fermion species in QUeST number four despite its four-dimension complex quaternion coordinates. It supports a 3+1 dimension Complex Quaternion Lorentz group. The cause is the central role of the speed of light c in QUeST (and in UST and MOST). We set c = 1. A massless particle travels at the speed of light in QUeST in the 24 spatial dimensions of QUeST, and satisfies $p^{02} - ||\mathbf{p}||^2 = 0$ where $||\mathbf{p}||$ is given by eq. 8.2 below. The speed of light separates the boosts of the Complex Quaternion Lorentz group into four types.

The four types of boosts in this 3+1 dimension space-time that boost a particle rest state to a state of motion with a real-valued energy[15] and a real-valued or complex quaternion spatial momentum are:[16]

1. A boost from rest to a frame with real-valued energy and momentum with $p^{02} - ||\mathbf{p}||^2 > 0$. A “normal charged lepton-like” fermion.

2. A boost from rest to a frame with real-valued energy and momentum with $p^{02} - ||\mathbf{p}||^2 < 0$. A “tachyonic neutral lepton-like” fermion.

---

[15] The energy must be real-valued to have a stable (in the absence of interactions) fundamental particle. In 3 + 1 real-valued space-time complex energy implies a “resonance” that decays.

[16] QUeST and MOST space-times must have a speed of light which we will denote as c. The speed of light distinguishes between “normal” and tachyon particles.

3. A boost from rest to a frame with real-valued energy and a complex quaternion valued spatial momentum with $p^{02} - ||\mathbf{p}||^2 > 0$. An "up-type quark-like" fermion.

4. A boost from rest to a frame with real-valued energy and a complex quaternion valued spatial momentum with $p^{02} - ||\mathbf{p}||^2 < 0$. A "tachyonic down-type quark-like" fermion.

where p^0 is the energy and $||\mathbf{p}||$ is the norm of the complex quaternion valued spatial 3-vector $\mathbf{p}$ with the form in 3+1-dimension complex quaternion space-time:

$$\mathbf{p} = \mathbf{p}_r + i\mathbf{p}_i + j\mathbf{p}_3 + k\mathbf{p}_4 + q\mathbf{p}_5 + r\mathbf{p}_6 + s\mathbf{p}_7 + t\mathbf{p}_8 \qquad (8.2)$$
$$||\mathbf{p}|| = \text{sqrt}(\mathbf{p}_r{\cdot}\mathbf{p}_r + \mathbf{p}_i{\cdot}\mathbf{p}_i + \mathbf{p}_3{\cdot}\mathbf{p}_3 + \ldots + \mathbf{p}_8{\cdot}\mathbf{p}_8)$$

where i, j, k, q, r, s, and t are fundamental quaternion units, and where each momentum component $\mathbf{p}_k$ is a 3-vector. The terms like $\mathbf{p}_r{\cdot}\mathbf{p}_r$ are 3-vector inner products.

In UST the momentum has the general form

$$\mathbf{p} = \mathbf{p}_r + i\mathbf{p}_i \qquad (8.3)$$

See Appendix A for details.

Normal and tachyonic particles are distinguished in 4 dimension complex quaternion space-time by the sign of $p^{02} - ||\mathbf{p}||^2$ in a manner similar to 4-dimension complex space-time. QUeST fermions occur in four species just like UST fermions.

8.4 MOST Fermion Species

In the Megaverse, the fermion species in MOST number four despite the 8-dimension complex quaternion nature of the space-time extracted from 32 dimension complex octonion space. The eight complex quaternion space-time coordinates support a 7+1 dimension Complex Quaternion Lorentz group.

There are four types of boosts in this 7+1 complex quaternion valued space-time that boost a particle rest state to a state of motion with a real energy, and a real-valued or complex quaternion valued momentum (spatial coordinates):

1. A boost from rest to a frame with real-valued energy and momentum with $p^{02} - ||\mathbf{p}||^2 > 0$. A "normal charged lepton-like" fermion.

2. A boost from rest to a frame with real-valued energy and momentum with $p^{02} - ||\mathbf{p}||^2 < 0$. A "tachyonic neutral lepton-like" fermion.

3. A boost from rest to a frame with real-valued energy and a complex quaternion valued spatial 7-momentum with $p^{02} - ||\mathbf{p}||^2 > 0$. An "up-type quark-like" fermion.

4. A boost from rest to a frame with real-valued energy and a complex quaternion valued spatial 7-momentum with $p^{02} - ||\mathbf{p}||^2 < 0$. A "tachyonic down-type quark-like" fermion.

where p^0 is the energy, and where each momentum component $\mathbf{p_k}$ is a spatial 7-vector:

$$\mathbf{p} = \mathbf{p_r} + i\mathbf{p_i} + j\mathbf{p_3} + k\mathbf{p_4} + q\mathbf{p_5} + r\mathbf{p_6} + s\mathbf{p_7} + t\mathbf{p_8} \tag{8.4}$$
$$||\mathbf{p}|| = \text{sqrt}(\mathbf{p_r}\cdot\mathbf{p_r} + \mathbf{p_i}\cdot\mathbf{p_i} + \mathbf{p_3}\cdot\mathbf{p_3} + \ldots + \mathbf{p_8}\cdot\mathbf{p_8})$$

where i, j, k, q, r, s, and t are fundamental quaternion units, and where $||\mathbf{p}||$ is the norm of $\mathbf{p}$. The terms like $\mathbf{p_r}\cdot\mathbf{p_r}$ are 7-vector inner products.

8.5 Transitions Between Fermions in Different Species

Fermions appear in species, in generations, and in layers. (See Fig. 1.7) Generation groups'[17] transformations can transform a given fermion to a corresponding fermion in another generation within the same layer. Layer groups'[18] transformations can transform a fermion to the corresponding fermion in a different layer (while

---

[17] Note there are eight Generationr groups: one for each layer for both the normal and Dark fermion sectors.

[18] Note there are eight Layer groups: one for each of the four generations for both the normal and Dark fermion sectors.

retaining the species and generation identity.) Dark groups'[19] transformations can transform between corresponding fermions in the normal and Dark fermion sectors.[20]

Transformations between fermions in different species in the same generation and layer require the use of SU(2)⊗U(1) and SU(3) as well as Complex Lorentz group transformations. See Fig. 8.1.

The role of each group in fermion transformations between species is:

1. SU(2) – transforms between e and ν, and between up-quark and down-quark based on the "charged" SU(2) generators.
2. SU(3) – transforms among the three up-quarks, and transforms among the three down-quarks.
3. Complex Lorentz Group – transforms among each of the four species using transformations in section 8.3 (and Appendix A). Complex Lorentz group transformations map among all four fermion species: Dirac ↔ tachyon ↔ up-quark ↔ down-quark.

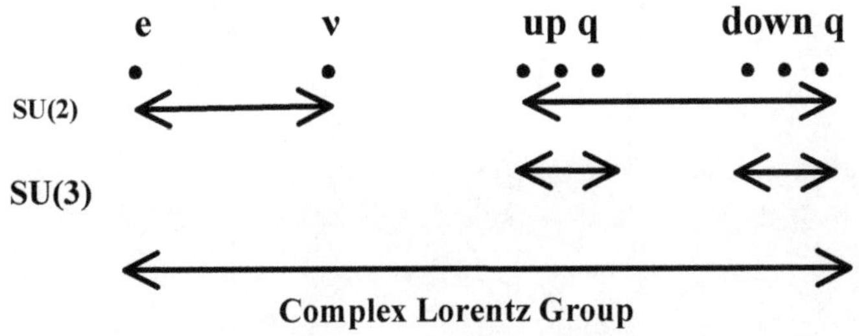

Figure 8.1. Transformations between fermions in differing species (and colors for quarks) for each generation in each layer.

The result of the set of all UST transformations (and similarly for QUeST transformations, which are analogous) is:

[19] Note there are four Dark groups: one for each of the four layers.

[20] We treat the set of fermions as free in these discussions.

Any fermion can be transformed to any other fermion.

There are no isolated fermions. In the 256 fermion UST and QUeST spectrums.

As a result all fermions can be generated from any one fermion by applying the above transformations. Bearing in mind dimension-fermion duality we can view the dimension dual of the one fermion as the "source" of fermions and corresponding dimensions with the other 255 particles and dimensions being generated by transformations parallel to the above SU(2)⊗U(1)⊗SU(3) and Complex Lorentz group transformations.

9. Particle Functionals Features

9.1 The Logic Core of Fundamental Fermions and Bosons

In previous books we opened the possibility that fermions (and bosons) might have a core that embodies logic in the form of spin as well as bare masses in the case of fermions. We defined functionals of various spins: 0, ½, 1, and 2. We saw that the core of spin ½ fermion functionals (that we called *qubes* in analogy with *qubits*) have a bare mass that we denoted m_0.

Bosons have cores as well that are boson functionals. Analogously we called a boson core a *quba*[21]. Boson functionals are massless. Bosons acquire masses through interactions.

The rationales for logic cores for particles is discussed in detail in chapters 3 and 8 of Blaha (2018e) and (2020c). We showed that the formalism based on a space of all particle functionals can lead to an explanation of the 'spooky' action at a distance of Quantum Entanglement that has been the subject of much discussion.

9.1.1 The Logic Building Block of Fermions – Qube Cores

If we consider all possible 'things' that might constitute a fundamental building block for a fundamental fermion theory they are all, at best, *ad hoc* and raise questions of their necessity and whether they are composed of yet a more fundamental substructure.

There is only one choice of building block that avoids these issues – a logic unit or qubit. A qubit is a fundamental entity that is a complex form of computer bit. A bit (and thus a qubit) is known to have an energy, or equivalently a mass, and has no

---

[21] We use 'quba' simply because of its similarity to 'qube'. The leading 'b' signifies its bosonic use. We pronounce 'quba' as 'bub' with a silent 'e.' The word 'quba', itself, is the name of a Bantu language spoken by the Bubi people of Bioko Island in Equatorial Guinea.

constituents of a more primitive form.[22] We call a unit of logic that forms the core of a particle a *qube*.[23] It exists as the core of a particle. But, in itself, it has no *independent* material existence or space-time coordinates. A qube is a functional that acquires features such as coordinates, through functional inner products to become an elementary particle. We define a qube as a fermion field theory functional. (See chapters 3 and 8 of Blaha (2018e) or (2020c).)

9.1.2 Mass of a Qube

Recent experiments have shown that a logical value of a qubit has an energy associated with it. One bit of information has about 3 x 10^{-21} joules of energy[24] or a rest mass, m_0, or about 0.02 eV using $m_0 = E/c^2$. This result was confirmed by E. Lutz et al.[25] who showed that there is a minimum amount of heat produced per bit of erased data. This minimal heat is called the *Landauer*[26] *limit*. The equivalent mass we will call the *Landauer mass* and denote it as m_0. We will assume that a fundamental Landauer mass exists in our discussions although the precise value of the mass will not be used since we may expect all physical particle masses to be renormalized to different values when interactions are taken into account.

We will assume all fermions contain a qube within them. (As stated above bosons do not have qubes within them. We call their core a quba.) A qube is assumed to have mass m_0. *The masses of fermions are modified to their known values by interactions.*

It is intriguing that the mass of the electron neutrino has been measured in a variety of experiments and found to be within an order of magnitude or so larger than our estimate of the Landauer mass (as we would expect since particles acquire a 'cloud

---

[22] A qube is a physical manifestation of a logical value. The relation of a qube to a logical value is analogous to the relation of a penciled point placed on paper to the concept of a point as a primitive in geometry.

[23] In the Blaha (2018f) we called qubes iotas. However, since the name iota was previously used as a particle name many years ago it seemed reasonable to use a different name. We chose the name 'qube' for self-evident reasons. *'Qube' is pronounced 'cube.'*

[24] E. Muneyuki et al, *Nature Physics*, DOI: 10.1038/NPHYS1821.

[25] E. Lutz et al, Nature **483** (7388): 187–190,10.1038/nature10872, (2012).

[26] R. Landauer, "Irreversibility and heat generation in the computing process", IBM Journal of Research and Development **5** (3): 183–191, (1961).

of virtual particles' due to interactions.) This 'cloud' can be expected to increase its mass above the Landauer mass. Since neutrinos only have the weak interaction it is not surprising that the increase due to interactions should not be large. The Mainz Neutrino Mass Experiment, for example, estimates the electron neutrino mass to be less than 2 eV. The new Karlsruhe Tritium Neutrino Experiment (September, 2019) found an upper limit of less than 1.1 eV.

A number of astronomical studies have also generated estimates of neutrino masses. In July 2010 the 3-D MegaZ DR7 galaxy survey found a limit for the combined mass of the three neutrino varieties to be less than 0.28 eV.[27] A smaller upper bound for the sum of neutrino masses, 0.23 eV, was found in March 2013 by the Planck collaboration,[28] In February 2014 a new estimate of the sum was found to be 0.320 ± 0.081 eV due to discrepancies between the Planck's measurements of the Cosmic Microwave Background, and other predictions, combined with the assumption that neutrinos are the cause of weaker gravitational lensing than implied by massless neutrinos.[29]

Thus the experimentally measured values of neutrino masses are consistent with the qube Landauer mass estimate of 0.02 eV given above. We thus assume that *a fermion particle consists of a qube with a certain mass,[30] which is renormalized, together with other features. These features will emerge later in the derivation of the complete theory.[31]*

We view Reality as ultimately a representation (or painting) of logic values evolving through interactions in time and space.[32]

---

[27] S. Thomas et al, "Upper Bound of 0.28 eV on Neutrino Masses from the Largest Photometric Redshift Survey", Physical Review Letters **105**: 031301 (2010).

[28] Planck Collaboration, arXiv:1303.5076 (2013).

[29] R. A. Battye et al, "Evidence for Massive Neutrinos from Cosmic Microwave Background and Lensing Observations", Phys. Rev. Lett. **112,** 051303 (2014).

[30] Leibniz first proposed the idea of logic 'particles' which he called monads. Our definition of a logic 'particle' does not include (or exclude) the presence of a spiritual part which was part of the definition of Leibniz's monads.

[31] A recent experiment claims to separate the spin part (which we identify as a logical value later) of a molecule from the rest of the molecule.

[32] Those who might suggest matter is substantial, and logic values are not, should remember that matter would be completely insubstantial if there were no forces in nature. Neutrinos which are close to insubstantial would be completely insubstantial if there were no weak interactions.

9.2 Quba Cores of Fundamental Bosons

We defined a corresponding boson functional quba for each type of elementary boson. We designated a boson functional as b_s where s specifies the spin which may be 0, 1, or 2. Every boson contains a boson functional core within it. A quba has the spin of the elementary boson within which it resides. It has zero mass since bosons are typically massless prior to symmetry breaking effects.

An important consequence of the masslessness of qubas is that they have no tachyon equivalents. Note: the bare mass of qubes led to tachyons. The masslessness of qubas prevents Complex Lorentz boosts from generating tachyonic bosons.

9.3 Particle Functional Space

The functionals of elementary particles, form a point space[33] that includes all the free field fermion and boson functionals of our universe and any other universe that might exist (the Megaverse). All fundamental fermions and bosons have a corresponding particle functional. Fermion particle functionals f… are labeled with momentum k, internal symmetry quantum numbers denoted λ, and spin (See eq. 6.2 for the factorized particle functional.) Boson particle functionals b… are similarly labeled with spin s, and internal symmetry quantum numbers.

9.4 Wave Space

We assume the space-time distance between Fourier wave function expansions to be *zero* in keeping with the zero distance between particle functionals in functional space. This assumption is solidly based on the instantaneity of transformations of parts of entangled states. (No spookiness!) Separating the parts of a quantum state S into space-like separated parts S_1 and S_2.we find a change in one part causes an instantaneous change in the other part:

$$\langle x|S\rangle \rightarrow \langle x_1|S_1\rangle\langle x_2||S_2\rangle \qquad (9.1)$$

irrespective of distance since the implicit functionals and Fourier expansions have no space-time separation from each other.

---

[33] Much of this chapter appears in the Blaha (2018a).

9.5 Skeleton Functional Lagrangians

If we could imagine a 'snapshot' of the universe[34] at one instant of time we could presumably enumerate all the functionals of the universe's particles. Then succeeding snapshots would show an ebb and flow of functionals as time progresses. This thought brings us to the important issue of the transformations of particle functionals in particle interactions. The simplest statement that one could make about functional transformations is that they are created and annihilated according to the interaction terms of the skeletonized Unified SuperStandard Theory (excluding quadratic terms which do not transform functionals.)

We skeletonize a lagrangian density by deleting all quadratic terms and replacing all particle fields by their corresponding functionals.[35] For example the lagrangian

$$\mathcal{L} = \bar{\psi}_C(i\gamma^\mu D_\mu - m)\psi_C(x) + b(\bar{\psi}_C\psi_C(x))^2 \qquad (9.2)$$

becomes the skeleton lagrangian

$$\mathcal{L}_S = bf^4 \qquad (9.3)$$

where f is the fermion's functional.

Thus our skeletonized lagrangian formalism describes the transitions between functionals in an interaction. This formalism is made more concrete by considering Feynman diagrams for the interactions.

---

[34] We realize that such a snapshot is not possible since infinite velocity particles that could feed a camera this snapshot do not exist.

[35] In our construction of particle functional space we have not introduced complex conjugation of functionals for lack of a good reason. Complex conjugation takes place only in the Fourier expansion part of a quantum field. Another issue is the appearance of lagrangian terms with factors that are derivatives of fields. Since we do not do computations with skeleton lagrangians we can ignore the derivative in each such factor and simply substitute the functional. For example, $\varphi^3(\partial^\mu\varphi)^2$ becomes the quba expression b^5.

9.6 Functional Interactions and Feynman Diagrams

Feynman diagrams with their in and out ordering specify the transformations between functionals more completely. A simple example shows the interaction transformations of functionals. Consider the lagrangian term

$$(\bar{\psi}\psi(x))^2(\partial^\mu\varphi)^2$$

A corresponding Feynman diagram for it is

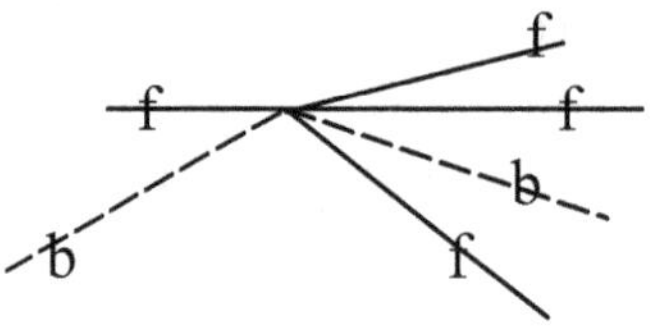

Figure 9.1.Functional Feynman diagram for the above interaction.

with qubes labeled f and qubas labeled b.

When internal symmetries are introduced then the skeletonized lagrangians and the corresponding Feynman diagram representations would be significantly more complicated.

9.7 Functional Space and Feynman Path Integrals

Functionals appear in Feynman Path Integrals and in Faddeev-Popov gauge fixing path integrals. We illustrate the use of functionals in the example:

$$Z(J) = N\int\prod_y dy \prod_\varphi d\varphi(y) \exp\{i\int d^4y[\mathscr{L}(\varphi(y) + J^\mu(y)\varphi(y)]\} \quad (9.4)$$

which in functional notation becomes?

$$Z(J) = N\int\prod_y d(y) \prod_b db \exp\{i\int d^4y[\mathscr{L}(\varphi(y) + J^\mu(y)\varphi(y)]\} \quad (9.5)$$

where (y) represents the Fourier expansion in the y coordinates, and with the implied inner product φ(y) = (b, (y)).

9.8 Functional Space and Feynman Path Integrals

Functionals appear in Feynman Path Integrals and in Faddeev-Popov gauge fixing path integrals. We illustrate the use of functionals in the example:

$$Z(J) = N\int \prod_{y} dy \prod_{\varphi} d\varphi(y) \exp\{i\int d^4y[\mathscr{L}(\varphi(y)) + J^{\mu}(y)\varphi(y)]\} \qquad (9.6)$$

which in functional notation becomes

$$Z(J) = N\int \prod_{y} d(y) \prod_{b} db \exp\{i\int d^4y[\mathscr{L}(\varphi(y)) + J^{\mu}(y)\varphi(y)]\} \qquad (9.7)$$

where (y) represents the Fourier expansion in the y coordinates, and with the implied inner product φ(y) = (b, (y)).

10. Monads and Observability

We have seen how particles appear to contain functionals that enable instantaneous quantum entanglement phenomena to take place without conflict with the Theory of Relativity. In this chapter we will see that observability is also directly understood within a monad framework.

Individual fundamental fermions and bosons contain functionals that we will call *monads* with certain physical properties. Fermions have qubes; bosons have qubas. We discussed their features earlier. Aggregates of fundamental particles form matter and energy. Thus matter and energy have monads within them.

10.1 Observability

We now turn to the questions of macroscopic observability and quantum observability. At both of these levels of observation monads play a decisive role. At the quantum level we have seen that monads enable instantaneous quantum phenomena to happen without conflict with Special Relativity. Quantum phenomena require observability when measured. *Thus monads, being intimately related to quantum phenomena, may also be viewed as the mechanism for observability at the quantum level.* This feature of monads was discussed in Blaha (2018f) and (2020c). Monads are then the instrument of observability at the quantum level of individual particles.

They are also the instrument of observability at the macroscopic level of aggregates of particles. Being ubiquitous monads answer questions of observability that are frequently posed, which are forms of the question:

Do events take place in the absence of observers?

Answer: YES, because the monads within particles provide a form of "default" level of observation.

10.2 Absolute Reality

Since all monads (functionals) exist in a space with no distance measure there is no question of disparities associated with space-time distances. Thus there is an *absolute reality* from instant to instant—all parts of the universe simultaneously exist *and are in contact with each other in principle.*[36] The universe has a unitary reality with all parts interrelated in the manner that thinkers have hypothesized for thousands of years. (For example: "The universe is one.") The events on a distant star, whose starlight we view, are as real as nearby events on earth. Aggregates of particles, which contain aggregates of monads, unfold dynamically according to physical law.

10.3 Relation to Consciousness

Consciousness has many facets. One facet is the reaction to events. In particle interactions monads transition (in general) to other monads. They do so in accord with quantum theory.

Monads thus have the general feature of reacting to events in a quantum probabilistic manner. They make "decisions" in accord with particle dynamics. Thus they have some features associated with consciousness:

1. Reaction to events

2. Selection of a path from the event point in a quantum probabilistic manner

Monads lack intelligence and decision making capacity beyond quantum dynamics. However many living creatures have similar abilities and limitations. Many creatures react to events according to a genetically determined pattern. One may characterize these creatures as partially conscious. In the next chapter we consider different levels of consciousness. The level of monad consciousness seems to be consistent with the levels of most living creatures. Mankind and other semi-intelligent species will be seen to have a higher level of consciousness.

---

[36] In particular the ubiquitous presence of gravitons—the essence of space-time—provides a universal connection mechanism as we discuss in chapter 11.

10.4 The Spark of Monads

Some thinkers (for example Leibniz) have attributed a spark of consciousness or spirit to monads This property is not determined by physical considerations and will not be considered here.

Appendix A. Lorentz Groups and the Four Species of Fermions From Blaha (2007b)

2. Lorentz Groups

2.0 Overview

This chapter discusses extensions of the Lorentz group of Special Relativity that includes transformations to reference frames moving at relative speeds greater than the speed of light in sections 2.1 through 2.4. We describe two groups that implement this extension: the Left-handed Extended Real Lorentz Group and the Right-handed Extended Real Lorentz Group. The difference in handedness is explored in detail in chapter 3 and ultimately leads to parity violation[37] in the Standard Model.

In section 2.5 we consider properties of the complexified Lorentz Group, L_C. In chapter 3 we show that L_c boosts provide three spin ½ alternatives to the Dirac equation. The four spin ½ free dynamical equations correspond to two lepton species and two quark species. The quark fields have an inherent global SU(3) symmetry that leads us to postulate quark color SU(3) (a local Yang-Mills interaction) with quarks in the fundamental representation $\underline{3}$ of SU(3).

2.1 The Lorentz Group

The Lorentz group, with which we are familiar, relates the coordinates of an event in two coordinate systems that differ by a relative velocity whose magnitude is less than the speed of light. The inhomogeneous Lorentz group includes coordinate displacements. We will discuss the relationship between coordinates modulo displacements, and not consider the displacement generators, since they will be the same for both ordinary Lorentz group transformations, and L_c transformations. Thus we will consider the homogeneous Lorentz group, and its generalizations that include faster-than-light transformations.

The elements of the Lorentz group, $\Lambda(\mathbf{v})$, when treated as 4×4 matrices satisfy

---

[37] Thus explaining the 50 year old question of the origin of parity violation discussed by T. D. Lee and C. N. Yang, Phys. Rev. **104**, 254 (1956). Parity violation originates in physics beyond the speed of light.

$$\Lambda(\mathbf{v})^{\mathrm{T}} G \Lambda(\mathbf{v}) = G \tag{2.1}$$

where G is the 4 x 4 diagonal matrix form of the metric diag(1, –1, –1, –1), the superscript "T" indicates the transpose, and **v** is the relative velocity of the reference frames. Note g^{00} is +1 in our metric. When $\Lambda(\mathbf{v})$ is a boost the components[38] of its matrix representation are real:

$$\Lambda(\mathbf{v}) = \begin{bmatrix} \gamma & -\gamma v_x & -\gamma v_y & -\gamma v_z \\ -\gamma v_x & 1 + (\gamma - 1)v_x^2/v^2 & (\gamma - 1)v_x v_y/v^2 & (\gamma - 1)v_x v_z/v^2 \\ -\gamma v_y & (\gamma - 1)v_x v_y/v^2 & 1 + (\gamma - 1)v_y^2/v^2 & (\gamma - 1)v_y v_z/v^2 \\ -\gamma v_z & (\gamma - 1)v_x v_z/v^2 & (\gamma - 1)v_y v_z/v^2 & 1 + (\gamma - 1)v_z^2/v^2 \end{bmatrix} \tag{2.2}$$

where $\gamma = (1 - v^2)^{-½}$, c = 1, $\mathbf{v} = (v_x, v_y, v_z)$, and $v = |\mathbf{v}|$. The boost transformation $\Lambda(\mathbf{v})$ can be expressed in the form

$$\Lambda(\mathbf{v}) = \exp[i\omega\hat{\mathbf{u}}\cdot\mathbf{K}] \tag{2.3}$$

where $\mathbf{v} = \hat{\mathbf{u}} \tanh(\omega)$, $\hat{\mathbf{u}}\bullet\hat{\mathbf{u}} = 1$, and **K** is the boost vector. Using the unit normalized velocity vector $\mathbf{u} = (u_x, u_y, u_z)$ the Lorentz boost matrix can be written:

$$\Lambda(\omega, \mathbf{u}) = \Lambda(\mathbf{v}) \tag{2.4}$$

---

[38] We shall consider only the proper, orthochronous Lorentz group until section 2.3 where we extend the discussion to the other subgroups of the Lorentz group and extended Lorentz groups.

$$= \begin{bmatrix} \cosh(\omega) & -\sinh(\omega)u_x & -\sinh(\omega)u_y & -\sinh(\omega)u_z \\ -\sinh(\omega)u_x & 1 + (\cosh(\omega) - 1)u_x^2 & (\cosh(\omega) - 1)u_x u_y & (\cosh(\omega) - 1)u_x u_z \\ -\sinh(\omega)u_y & (\cosh(\omega) - 1)u_x u_y & 1 + (\cosh(\omega) - 1)u_y^2 & (\cosh(\omega) - 1)u_y u_z \\ -\sinh(\omega)u_z & (\cosh(\omega) - 1)u_x u_z & (\cosh(\omega) - 1)u_y u_z & 1 + (\cosh(\omega) - 1)u_z^2 \end{bmatrix}$$

This definition of the general form of the proper, orthochronous, Lorentz boost matrix $\Lambda(\omega, \mathbf{u})$ will be used in subsequent sections to define extended faster-than-light boost transformations.

The vector form of a Lorentz boost transformation is

$$\begin{aligned} \mathbf{x}' &= \mathbf{x} + (\gamma - 1)\mathbf{x}\cdot\mathbf{v}\ \mathbf{v}/v^2 - \gamma\mathbf{v}t \\ t' &= \gamma(t - \mathbf{v}\cdot\mathbf{x}/c^2) \end{aligned} \tag{2.5}$$

where $\gamma = (1 - \beta^2)^{-1/2}$ with $\beta = v/c = v$ (with $c = 1$).

2.2 Sets of Real Matrices $\Lambda(\omega, u)$ for Complex ω

Faster-than-light transformations will be implemented with specific complex forms of ω as we will see in the next section. Since

$$\cosh^2(z) - \sinh^2(z) = 1 \tag{2.6}$$

for any complex values z it is easy to show that

$$\Lambda(\omega, \mathbf{u})^T G \Lambda(\omega, \mathbf{u}) = G \tag{2.7}$$

for any complex ω. Eq. 2.7 implies $\Lambda(\omega, \mathbf{u})$ is a member of the Lorentz group or of the complex Lorentz group for complex ω.

For certain values of the imaginary part of ω the matrix $\Lambda(\omega, \mathbf{u})$ has a simple form, similar to that of $\Lambda(\omega, \mathbf{u})$ for real ω, but which generates boosts to relative speeds greater than the speed of light. Among these values are:

$$\omega \rightarrow \omega_{\pm} = \omega \pm i\pi/2 \tag{2.8}$$

In chapter 3 (cases 1 and 2 in section 3.7) we will see that the alternate choices $\omega_{\pm}$ correspond to specific physical situations when considered within the framework of the L_c group.

2.3 Extensions of the Lorentz Group to Faster-than-Light Transformations

In this section we will substitute $\omega_{\pm}$ for ω in $\Lambda(\omega, \mathbf{u})$ and then show that we obtain two sets of possible transformations to faster-than-light reference frames. One set of transformations where $\omega_L = \omega + i\pi/2$ will be called *left-handed superluminal boosts.* When they are combined with the Lorentz group we are led to the "left-handed" Standard Model. We denote members of this set, $\Lambda_L(\omega, \mathbf{u})$, with the subscript "L".

The other set of boosts where $\omega_R = \omega - i\pi/2$ will be called *right-handed superluminal boosts*. When they are combined with the Lorentz group they lead to a right-handed, unphysical, version of the Standard Model. We denote members of this set, $\Lambda_R(\omega, \mathbf{u})$, with the subscript "R".

Before considering faster-than-light boosts we note the relation between ω in a Lorentz boost $\Lambda(\omega, \mathbf{u})$, and the magnitude of the relative velocity $v < 1$, is

$$\mathbf{v} = \hat{\mathbf{u}} \tanh(\omega) \qquad\qquad \hat{\mathbf{u}}\cdot\hat{\mathbf{u}} = 1$$

$$\begin{aligned} \cosh(\omega) &= \gamma = (1 - v^2)^{-1/2} \\ \sinh(\omega) &= v\gamma = \beta\gamma \end{aligned} \tag{2.9}$$

where $\beta = v = |\mathbf{v}|$.

Left-Handed Superluminal Transformations

Left-handed (proper orthochronous) superluminal boost transformations $\Lambda_L(\mathbf{v})$ have the same form as eq. 2.2 for ordinary (proper orthochronous) Lorentz boost transformations. However the magnitude of the relative velocity **v** is greater than the speed of light. Thus $\gamma = (1 - v^2)^{-½}$ is pure imaginary and $\Lambda_L(\mathbf{v})$ is complex.

$$\Lambda_L(\mathbf{v}) = \begin{bmatrix} \gamma & -\gamma v_x & -\gamma v_y & -\gamma v_z \\ -\gamma v_x & 1 + (\gamma - 1)v_x^2/v^2 & (\gamma - 1)v_x v_y/v^2 & (\gamma - 1)v_x v_z/v^2 \\ -\gamma v_y & (\gamma - 1)v_x v_y/v^2 & 1 + (\gamma - 1)v_y^2/v^2 & (\gamma - 1)v_y v_z/v^2 \\ -\gamma v_z & (\gamma - 1)v_x v_z/v^2 & (\gamma - 1)v_y v_z/v^2 & 1 + (\gamma - 1)v_z^2/v^2 \end{bmatrix} \qquad (2.10)$$

This transformation raises several issues – the most prominent of which is the interpretation of the imaginary coordinates generated by the transformation. Imaginary coordinates would appear at first glance to be unphysical. However if we view the measurement of these quantities operationally: an observer measures distances with "rulers", and time with clocks, which both give real numeric values. Thus an observer in a coordinate system with imaginary coordinates will be "unaware" of the imaginary nature of these quantities. It is only when the observer's coordinate system is related to another coordinate system through a superluminal transformation as done above that the imaginary nature of the coordinates becomes evident. From this point of view imaginary coordinates pose no new conceptual issues.

It will become apparent that a related transformation defined by

$$E(\mathbf{v}) = i\Lambda_L(\mathbf{v}) \qquad (2.10a)$$

should be used to define the coordinates generated by a superluminal boost transformation.[39] This definition of coordinates leads to the tachyonic Dirac equation for spin ½ tachyons. Therefore we define

$$X' = E(\mathbf{v})X = i\Lambda_L(\mathbf{v})X \qquad (2.10b)$$

where X' and X are coordinates in the prime and unprimed reference frames respectively. X' and X are represented as column vectors.[40] If we consider the simple case of a relative velocity v in the x direction, then the E boost gives

$$\begin{aligned} t' &= |\gamma|(t - \beta x) \\ x' &= |\gamma|(x - \beta t) \\ y' &= iy \\ z' &= iz \end{aligned}$$

where $|\gamma|$ is the absolute value of γ. Thus y' and z' are imaginary from the viewpoint of the unprimed coordinate system.

Eqs. 2.7 and 2.10a imply

$$E(\omega, \mathbf{u})^T G E(\omega, \mathbf{u}) = -G \qquad (2.10c)$$

where $E(\omega, \mathbf{u}) \equiv E(\mathbf{v})$. Thus E inverts the sign of the metric tensor. As a result the invariant scalar product relation (in matrix form) is

$$XGY = X'(-G)Y' \qquad (2.10d)$$

[39] E is an upper case epsilon chosen to stand for Extended since the boosts in question are extended boosts (v > 1) that are not part of the Lorentz group.

[40] This definition of superluminal coordinate transformations is physically acceptable as long as allowance is made for the change in the definition of the invariant interval based on eq. 2.10c.

The metric tensor is represented by G. Most of the succeeding discussion will be phrased in terms of $\Lambda_L(\mathbf{v})$. Eq. 2.10a enables one to easily rephrase the discussion in terms of $E(\omega, \mathbf{u})$.

The Cosh-Sinh Representation of Left-Handed Superluminal Boosts

We will now develop the representation of left-handed superluminal boost transformations in terms of cosh(ω) and sinh(ω) for later use in our discussion of tachyons. We find that we must use a complex $\omega = \omega_L = \omega + i\pi/2$ to properly describe left-handed superluminal boosts. The relation between ω_L and v is different from eq. 2.9 for the case of left-handed superluminal boosts:

$$\begin{aligned}\cosh(\omega_L) &= i\sinh(\omega) = -\gamma = i\gamma_s \\ \sinh(\omega_L) &= i\cosh(\omega) = -\beta\gamma = i\beta\gamma_s\end{aligned} \tag{2.11}$$

where $\beta = v > 1$ and $\boldsymbol{\omega \geq 0}$. Note

$$\begin{aligned}\sinh(\omega) &= \gamma_s \\ \cosh(\omega) &= \beta\gamma_s\end{aligned} \tag{2.12}$$

with

$$\gamma_s = (\beta^2 - 1)^{-1/2} \tag{2.13}$$

Upon substituting ω_L for ω in eq. 2.4 we obtain another form for a left-handed superluminal transformation (equivalent to that of eq. 2.10):

$$\Lambda_L(\omega, \mathbf{u}) = \Lambda(\omega + i\pi/2, \mathbf{u}) = -iE(\omega, \mathbf{u})$$

$$= \begin{bmatrix} \cosh(\omega_L) & -\sinh(\omega_L)u_x & -\sinh(\omega_L)u_y & -\sinh(\omega_L)u_z \\ -\sinh(\omega_L)u_x & 1+(\cosh(\omega_L)-1)u_x^2 & (\cosh(\omega_L)-1)u_xu_y & (\cosh(\omega_L)-1)u_xu_z \\ -\sinh(\omega_L)u_y & (\cosh(\omega_L)-1)u_xu_y & 1+(\cosh(\omega_L)-1)u_y^2 & (\cosh(\omega_L)-1)u_yu_z \\ -\sinh(\omega_L)u_z & (\cosh(\omega_L)-1)u_xu_z & (\cosh(\omega_L)-1)u_yu_z & 1+(\cosh(\omega_L)-1)u_z^2 \end{bmatrix}$$

$$= \begin{bmatrix} i\gamma_s & -i\beta\gamma_s u_x & -i\beta\gamma_s u_y & -i\beta\gamma_s u_z \\ -i\beta\gamma_s u_x & 1 + (i\gamma_s - 1)u_x^2 & (i\gamma_s - 1)u_x u_y & (i\gamma_s - 1)u_x u_z \\ -i\beta\gamma_s u_y & (i\gamma_s - 1)u_x u_y & 1 + (i\gamma_s - 1)u_y^2 & (i\gamma_s - 1)u_y u_z \\ -i\beta\gamma_s u_z & (i\gamma_s - 1)u_x u_z & (i\gamma_s - 1)u_y u_z & 1 + (i\gamma_s - 1)u_z^2 \end{bmatrix} = \Lambda_L(\mathbf{v}) \qquad (2.14)$$

by eq. 2.10.

A simple case that illustrates the nature of the left-handed superluminal boost is to assume the relative velocity is in the x direction. Then eq. 2.14 becomes

$$\Lambda_L(\omega, \mathbf{u} = (1,0,0)) = \begin{bmatrix} i\gamma_s & -i\beta\gamma_s & 0 & 0 \\ -i\beta\gamma_s & i\gamma_s & 0 & 0 \\ 0 & 0 & 1 & 0 \\ 0 & 0 & 0 & 1 \end{bmatrix} \qquad (2.15)$$

implementing the coordinate transformation:

$$X' = -i\Lambda_L(\omega, \mathbf{u} = (1,0,0))X = E(\omega, \mathbf{u} = (1,0,0))X$$

or

$$\begin{aligned} t' &= \gamma_s(t - \beta x) \\ x' &= \gamma_s(x - \beta t) \\ y' &= iy \\ z' &= iz \end{aligned} \qquad (2.16)$$

The addition rule for the x-component of velocity can be computed for infinitesimal displacements in space and time:

$$\begin{aligned} v_x' = \Delta x' / \Delta t' &= (\Delta x\, \gamma_s - \Delta t\, \beta\gamma_s)/(\Delta t\, \gamma_s - \Delta x\, \beta\gamma_s) \\ &= (v_x - \beta)/(1 - \beta v_x) \end{aligned} \qquad (2.17)$$

(where $\beta = u$ is the relative speed) in the limit $\Delta t \to 0$ where the x component of a particle's velocity in the unprimed frame is $v_x = \Delta x/\Delta t$. $\Delta t'$ is determined by

$$\Delta t' = \Delta t\, \gamma_s(1 - \beta v_x) \tag{2.18}$$

Note the velocity of light is the same in the primed and unprimed reference frames. (If $v_x = 1$ then $v_x' = 1$.) *Thus left-handed superluminal transformations preserve the constancy of the speed of light in all reference frames.*

Further note that increasing the value of ω in $\Lambda_L(\omega, \mathbf{u})$ corresponds to decreasing the magnitude of the relative velocity v since

$$v = \text{cotanh}(\omega) \tag{2.19}$$

Thus when $v = 1$ then $\omega = \infty$, and when $\omega = 0$ then $v = \infty$.

General Velocity Transformation Law – Left-Handed Superluminal Boosts

The general velocity transformation law for a particle moving with velocity **v** in the unprimed reference frame and velocity **v'** in the primed reference frame is

$$\mathbf{v}' = \mathbf{w} + (\gamma - 1)\mathbf{w}\cdot\mathbf{v}\ \mathbf{w}/w^2 - \gamma\mathbf{w} \tag{2.20}$$

where **w** is the relative velocity of the primed reference frame with respect to the unprimed reference frame, and $\gamma = (1 - w^2)^{-1/2}$. Eq. 2.20 is obtained by calculating the derivative $d\mathbf{x}'/dt'$ using eqs. 2.5. The relative velocity **w** can be greater or less than the speed of light. Eq. 2.20 implies

$$v'^2 = 1 + (v^2 - 1)(1 - w^2)/(1 - \mathbf{w}\cdot\mathbf{v})^2 \tag{2.21}$$

The velocity transformation law can be used to determine the multiplication rules for Lorentz and extended Lorentz transformations (next subsection).

Left-Handed Transformations Multiplication Rules

In this subsection we will determine the multiplication rules of left-handed extended Lorentz boosts and show the Lorentz group is an invariant subgroup of the extended left-handed Lorentz group. To do this we will consider three reference frames: an unprimed frame, a "primed" frame moving with velocity **w** with respect to the unprimed frame, and a "double-primed" frame moving with velocity **v** with respect to the unprimed frame and velocity **v'** with respect to the primed frame. See Fig. 2.1.

A moment's consideration reveals that **v'** is related to **v** by eqs. 2.20 and 2.21. (Think of the double-primed coordinate system as a particle or attached to a particle. In addition note that the transformation law from the unprimed to the double-primed reference frame must be the product of consecutive transformations (boosts) from the unprimed to the primed reference frames and then from the primed to the double-primed reference frames:

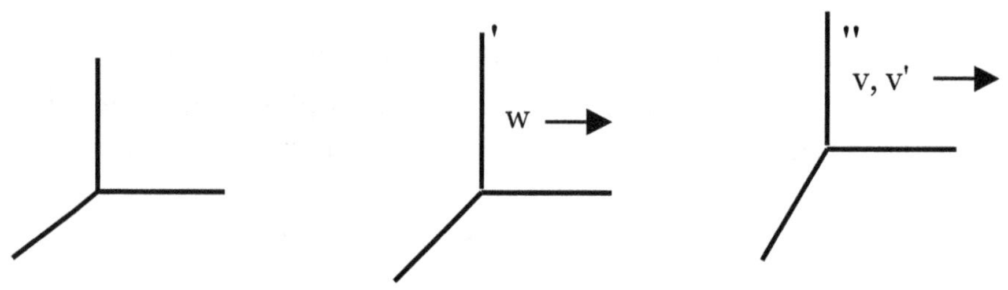

Figure 2.1. Three reference frames used to establish transformation multiplication rules.

$$\Lambda_?(\mathbf{v}) = \Lambda_?(\mathbf{v'})\Lambda_?(\mathbf{w}) \qquad (2.22)$$

where the "?" subscripts indicate Lorentz or superluminal transformations (boosts) depending on the magnitude of the relative velocity in the transformation's parentheses.

We now consider the cases using eq. 2.21:

1) If w > 1 and v' > 1
   then eq. 2.21 implies v < 1 and thus $\Lambda_?(\mathbf{v})$ is a Lorentz transformation

$$\Lambda(\mathbf{v}) = \Lambda_L(\mathbf{v'})\Lambda_L(\mathbf{w}) \tag{2.23}$$

2) If w > 1, v' < 1
   then eq. 2.21 implies v > 1 and thus $\Lambda_?(\mathbf{v})$ is a superluminal transformation

$$\Lambda_L(\mathbf{v}) = \Lambda(\mathbf{v'})\Lambda_L(\mathbf{w}) \tag{2.24}$$

3) If w < 1, v' > 1
   then eq. 2.21 implies v > 1 and thus $\Lambda_?(\mathbf{v})$ is a superluminal transformation

$$\Lambda_L(\mathbf{v}) = \Lambda_L(\mathbf{v'})\Lambda(\mathbf{w}) \tag{2.25}$$

4) If w < 1, v' < 1
   then eq. 2.21 implies v < 1 and thus $\Lambda_?(\mathbf{v})$ is a Lorentz transformation

$$\Lambda(\mathbf{v}) = \Lambda(\mathbf{v'})\Lambda(\mathbf{w}) \tag{2.26}$$

where, in each above case, the transformation on the left side of the equation may be a boost or a combination of a boost and a spatial rotation. Thus we have obtained the multiplication rules for left-handed extended Lorentz transformations.

The inverse of a Lorentz boost (eq. 2.3) is

$$\Lambda^{-1}(\omega, \hat{\mathbf{u}}) = \exp[-i\omega\hat{\mathbf{u}}\cdot\mathbf{K}] \tag{2.27}$$

where $\boldsymbol{\omega} \geq \mathbf{0}$. It shows the inverse is generated by letting $\omega \rightarrow -\omega$. Note that since v = tanhω, the effect of $\omega \rightarrow -\omega$ is to let v → –v. In the case of superluminal left-handed boosts, since

$$\Lambda_L(\omega, \mathbf{u}) = \Lambda(\omega + i\pi/2, \mathbf{u}) = \exp[i(\omega + i\pi/2)\hat{\mathbf{u}}\cdot\mathbf{K}] \quad (2.28)$$

we find the inverse is

$$\Lambda_L^{-1}(\omega, \mathbf{u}) = \Lambda(-(\omega + i\pi/2), \mathbf{u}) = \exp[-i(\omega + i\pi/2)\hat{\mathbf{u}}\cdot\mathbf{K}] \quad (2.29)$$

where $\boldsymbol{\omega} \geq \mathbf{0}$. Since $\Lambda_L^{-1}(\omega, \mathbf{u})$ is not the hermitean conjugate of $\Lambda_L(\omega, \mathbf{u})$, superluminal boosts are not unitary. However unitarity is not required since even complex Lorentz group elements satisfy the defining relation of the Lorentz group (eqs. 2.1 and 2.7). The effect of letting $\omega_L = \omega + i\pi/2 \rightarrow -(\omega + i\pi/2)$ is to let $v \rightarrow -v$ since

$$\tanh(\omega_L) = \beta = v \quad (2.30)$$

by eqs. 2.11.

We now turn to proving the Lorentz group is an invariant subgroup of the left-handed extended Lorentz group.

Theorem: The Lorentz group is an invariant subgroup of the left-handed extended Lorentz group.

Proof:

Consider the quantity

$$J = \Lambda_L(\omega_u, \mathbf{u})^{-1}\Lambda(\omega_v, \mathbf{v})\Lambda_L(\omega_u, \mathbf{u})$$

then

$$J = \Lambda_L(\omega_u, \mathbf{u})^{-1}\Lambda_L(\omega_z, \mathbf{z})$$
$$= \Lambda(\omega_w, \mathbf{w})$$

by eqs. 2.24 and 2.23 respectively for some ω_z, $\mathbf{z}$, ω_w and $\mathbf{w}$. Thus the Lorenz group is an invariant subgroup of the left-handed extended Lorentz group.

Right-Handed Superluminal Transformations

When we transform between reference frames using a right-handed[41] superluminal boost (where the magnitude of the relative velocity v is greater than c) the relation between ω and v also changes. The variable ω becomes $\omega_R = \omega - i\pi/2$ and

$$\cosh(\omega_R) = -i\sinh(\omega) = \gamma = -i\gamma_s \tag{2.31}$$
$$\sinh(\omega_R) = -i\cosh(\omega) = \beta\gamma = -i\beta\gamma_s \tag{2.32}$$

where $\beta = v > 1$ and $\boldsymbol{\omega \geq 0}$. Note (as before)

$$\sinh(\omega) = \gamma_s \tag{2.12}$$
$$\cosh(\omega) = \beta\gamma_s$$

with

$$\gamma_s = (\beta^2 - 1)^{-1/2} \tag{2.13}$$

Upon substituting ω_R for ω in eq. 2.4 we obtain the form of the right-handed superluminal boost and a corresponding transformation E_R:[42]

$$\Lambda_R(\omega, \mathbf{u}) = \Lambda(\omega - i\pi/2, \mathbf{u}) = -iE_R(\omega, \mathbf{u})$$

$$= \begin{bmatrix} -i\gamma_s & i\beta\gamma_s u_x & i\beta\gamma_s u_y & i\beta\gamma_s u_z \\ i\beta\gamma_s u_x & 1 + (-i\gamma_s - 1)u_x^2 & (-i\gamma_s - 1)u_x u_y & (-i\gamma_s - 1)u_x u_z \\ i\beta\gamma_s u_y & (-i\gamma_s - 1)u_x u_y & 1 + (-i\gamma_s - 1)u_y^2 & (-i\gamma_s - 1)u_y u_z \\ i\beta\gamma_s u_z & (-i\gamma_s - 1)u_x u_z & (-i\gamma_s - 1)u_y u_z & 1 + (-i\gamma_s - 1)u_z^2 \end{bmatrix} \tag{2.33}$$

---

[41] We call these transformations right-handed because they lead eventually to an alternate Standard Model with right-handed SU(2) doublets and left-handed SU(2) singlets. This alternate right-handed Standard Model does not appear to correspond to experimental reality.

[42] We note the singularities at $\beta = \pm 1$ or $\omega = \pm\infty$. As a result we have a branch cut in the complex ω-plane consisting of the entire real ω axis. Therefore three left-handed boosts are not equivalent to a right-handed boost but rather appear on another Riemann sheet.

A simple case that illustrates the nature of the right-handed superluminal transformation is to assume a relative velocity in the x direction. Then eq. 2.33 becomes

$$\Lambda_R(\omega, \mathbf{u} = (1,0,0)) = \begin{bmatrix} -i\gamma_s & i\beta\gamma_s & 0 & 0 \\ i\beta\gamma_s & -i\gamma_s & 0 & 0 \\ 0 & 0 & 1 & 0 \\ 0 & 0 & 0 & 1 \end{bmatrix} \quad (2.34)$$

implementing the coordinate transformation:

$$X' = E_R(\omega, \mathbf{u})X = i\Lambda_R(\omega, \mathbf{u})X$$

or

$$\begin{aligned} t' &= \gamma_s(t - \beta x) \\ x' &= \gamma_s(x - \beta t) \\ y' &= iy \\ z' &= iz \end{aligned} \quad (2.35)$$

Comparing eq. 2.34 with eq. 2.15 for a left-handed superluminal boost we see that

$$PT\Lambda_L(\omega, \mathbf{u} = (1,0,0)) = \begin{bmatrix} -i\gamma_s & i\beta\gamma_s & 0 & 0 \\ i\beta\gamma_s & -i\gamma_s & 0 & 0 \\ 0 & 0 & -1 & 0 \\ 0 & 0 & 0 & -1 \end{bmatrix} \quad (2.36)$$

where P is the parity operator and T is the time reversal operator. If we now apply a spatial rotation $\mathfrak{R}$ of π radians around the x axis then we obtain

$$\Re PT\Lambda_L(\omega, \mathbf{u} = (1,0,0))\Re^{-1} = \begin{bmatrix} -i\gamma_s & i\beta\gamma_s & 0 & 0 \\ i\beta\gamma_s & -i\gamma_s & 0 & 0 \\ 0 & 0 & 1 & 0 \\ 0 & 0 & 0 & 1 \end{bmatrix} \quad (2.37)$$

$$= \Lambda_R(\omega, \mathbf{u} = (1,0,0))$$

by eq. 2.34. Since P and T commute with spatial rotations we find

$$\Lambda_R(\omega, \mathbf{u} = (1,0,0)) = PT\Re\Lambda_L(\omega, \mathbf{u} = (1,0,0))\Re^{-1} \quad (2.38)$$

or, more generally, performing additional spatial rotations:

$$\Lambda_R(\omega, \mathbf{u}) = PT\Re_u\Re\Re_w\Lambda_L(\omega, \mathbf{w})\Re_w^{-1}\Re^{-1}\Re_u^{-1} \quad (2.39)$$

or,

$$\Lambda_R(\omega, \mathbf{u}) = PT\Re_{tot}\Lambda_L(\omega, \mathbf{w})\Re_{tot}^{-1} \quad (2.40)$$

where **u** and **w** are unit vectors. Alternately,

$$\Lambda_L(\omega, \mathbf{w}) = PT\Re_{tot}^{-1}\Lambda_R(\omega, \mathbf{u})\Re_{tot} \quad (2.41a)$$

or

$$\Lambda_L(\omega, \mathbf{w}) = PT\Lambda_R(\omega, \mathbf{u}') \quad (2.41b)$$

for some unit vector **u'**.

Thus we have shown that PT can be used to relate left-handed and right-handed boosts in a one-to-one fashion. *The appearance of the parity operator P takes on great significance when we derive features of the Standard Model. The appearance of left-*

handed doublets and right-handed singlets stems directly from the implicit parity dependence of the left-handed extended Lorentz group.

For the right-handed boost of eq. 2.34 the addition rule for the x-component of velocity can be computed for infinitesimal displacements in space and time:

$$v_x' = \Delta x'/\Delta t' = (\Delta x\, \gamma_s - \Delta t\, \beta\gamma_s)/(\Delta t\, \gamma_s - \Delta x\, \beta\gamma_s)$$
$$= (v_x - \beta)/(1 - \beta v_x) \qquad (2.42)$$

in the limit $\Delta t \to 0$ where the x component of a particle's velocity in the unprimed frame is $v_x = \Delta x/\Delta t$. Note if $v_x = 1$ then $v_x' = 1$. *Thus right-handed superluminal transformations also preserve the constancy of the speed of light in all reference frames.*

2.4 Inhomogeneous Left-Handed Extended Lorentz Group

The *Left-Handed Extended Lorentz group* consists of the elements of the Lorentz group plus left-handed superluminal transformations including pure boosts and combinations of boosts and spatial rotations. Thus homogeneous left-handed superluminal transformations have the general form:

$$\Lambda_L(\mathbf{v}, \boldsymbol{\theta}) = \exp[i\omega_L \hat{\mathbf{u}}\cdot\mathbf{K} + i\boldsymbol{\theta}\cdot\mathbf{J}] \qquad (2.43)$$

where $\omega_L' = \omega + i\pi/2$, $\boldsymbol{\theta}$ is the angular vector, and $\mathbf{J}$ is the angular momentum operator vector. Inhomogeneous left-handed superluminal transformations, which include displacements, can be expressed as

$$\Lambda_L(\mathbf{v}, \boldsymbol{\theta}, \mathbf{d}) = \exp[i\omega_L \hat{\mathbf{u}}\cdot\mathbf{K} + i\boldsymbol{\theta}\cdot\mathbf{J} - i\mathbf{d}\cdot\mathbf{P}] \qquad (2.44)$$

where $\mathbf{P}$ is the momentum operator vector and $\mathbf{d}$ is a displacement vector.

We have verified the group structure of the left-handed extended Lorentz group is indeed a group in the preceding section (eqs. 2.23 – 2.29).

We also note

$$\det \Lambda_L(\omega, \mathbf{u}) = \pm 1 \tag{2.45a}$$

by eq. 2.7.

The ordinary Lorentz group is divided into four disjoint subgroups that are often denoted:

$$\begin{aligned}
L_+^\uparrow&: \det \Lambda(\omega, \mathbf{u}) = +1;\ \operatorname{sgn} \Lambda(\omega, \mathbf{u})^0{}_0 = +1 \\
L_-^\uparrow&: \det \Lambda(\omega, \mathbf{u}) = -1;\ \operatorname{sgn} \Lambda(\omega, \mathbf{u})^0{}_0 = +1 \\
L_+^\downarrow&: \det \Lambda(\omega, \mathbf{u}) = +1;\ \operatorname{sgn} \Lambda(\omega, \mathbf{u})^0{}_0 = -1 \\
L_-^\downarrow&: \det \Lambda(\omega, \mathbf{u}) = -1;\ \operatorname{sgn} \Lambda(\omega, \mathbf{u})^0{}_0 = -1
\end{aligned} \tag{2.45b}$$

where $\operatorname{sgn} \Lambda(\omega, \mathbf{u})^0{}_0$ is the sign of the 00 component of the $\Lambda(\omega, \mathbf{u})$ matrix. The various subgroups are related by the discrete transformations of parity P and time reversal T:

$$L_+^\uparrow \xrightarrow{P} L_-^\uparrow$$

$$L_+^\uparrow \xrightarrow{PT} L_+^\downarrow$$

$$L_+^\uparrow \xrightarrow{T} L_-^\downarrow$$

The left-handed superluminal transformations are disjoint in a somewhat different way. By eq. 2.45a the determinants are ±1. However the 0-0 matrix element of eq. 2.7 gives

$$\Lambda_L{}^0{}_0{}^2 - \Sigma_i\, (\Lambda_L{}^i{}_0)^2 = 1 \tag{2.46}$$

The representation of superluminal boosts (eq. 2.14) shows that each factor in eq. 2.46 is imaginary. Thus eq. 2.46 implies

$$\Sigma_i |\Lambda_L{}^i{}_0|^2 \geq 1 \tag{2.47}$$

$$|\Lambda_L{}^0{}_0| \geq 0 \qquad \text{(not } \geq 1 \text{ as in eq. 2.45b)} \tag{2.48}$$

where || indicates absolute value. Thus the magnitude of $\Lambda_L{}^0{}_0$ does not have a gap. Therefore left-handed superluminal transformations can be divided into two categories:

$$\begin{aligned} {}_L L_+: &\quad \det \Lambda_L(\omega, \mathbf{u}) = +1 \\ {}_L L_-: &\quad \det \Lambda_L(\omega, \mathbf{u}) = -1 \end{aligned} \tag{2.49}$$

Under a PT transformation a left-handed superluminal transformation becomes a right handed superluminal transformation (eq. 2.41b).

Again the various disjoint pieces are related by the discrete transformations of parity P and time reversal T:

$$\begin{aligned} {}_L L_+ &\xrightarrow{P} {}_L L_- \\ {}_L L_+ &\xrightarrow{T} {}_L L_- \end{aligned} \tag{2.50}$$

2.5 Inhomogeneous Right-Handed Extended Lorentz Group

The inhomogeneous right-handed extended Lorentz group[43] consists of the Lorentz group plus right-handed superluminal transformations that have the form:

---

[43] Since γ_s has branch points at $v = \pm 1$ which corresponds to $\omega = \pm\infty$ for the left-handed and right-handed groups there is a cut along the real ω axis between $-\infty$ and $+\infty$. Therefore the product of three left-handed Lorentz transformations does not yield a right-handed transformation (as might be supposed from eqs. 2.43 and 2.51) but rather a left-handed transformation on the second sheet. A transformation with $\omega + 3i\pi/2$ is not equivalent to any with $\omega - i\pi/2$.

$$\Lambda_R(\mathbf{v}, \boldsymbol{\theta}, \mathbf{d}) = \exp[i\omega_R\hat{\mathbf{u}}\cdot\mathbf{K} + i\boldsymbol{\theta}\cdot\mathbf{J} - i\mathbf{d}\cdot\mathbf{P}] \quad (2.51)$$

in general where $\omega_R = \omega - i\pi/2$.

2.6 The Complex Lorentz Group L(C)

The complex Lorentz group consists of all complex transformations of a 4-dimensional complex space-time that satisfies eq. 2.1 when the transformations are expressed as 4 by 4 matrices composed of complex numbers. At first glance the complex Lorentz group would not appear to be relevant to our universe since our universe seems to be described by real space-time coordinates. However as Streater (1964) points out,[44] "It is essential in the proof of the PCT theorem." Thus the complex Lorentz group has a deep role in Quantum Field Theory. The real homogeneous Lorentz group has six parameters – three boost parameters and three rotation parameters. The homogeneous complex Lorentz group has 12 generators and 12 parameters – six boost parameters and six rotation parameters.

SL(2,C)⊗SL(2,C) Formulation of Complex Lorentz Group

The complex Lorentz group can be realized in matrix form as a map from the proper orthochronous Lorentz group $L_+^\uparrow$ to SL(2,C):

$$\Lambda(A) \rightarrow A \quad (2.52)$$

where

$$A = A^0 I + \mathbf{A}\cdot\boldsymbol{\tau} \quad (2.53)$$

with $\boldsymbol{\tau}$ a 3-vector of the 2 × 2 Pauli matrices. The four coefficients A^0 and $\mathbf{A}$ are *complex* numbers. A has determinant one. The Lorentz transformation

$$x' = \Lambda(A)x \quad (2.54)$$

corresponds to the transformation

---

[44] Streater (1964) p. 13.

$$x'^0 I + \mathbf{x}'\cdot\boldsymbol{\tau} = A(x^0 I + \mathbf{x}\cdot\boldsymbol{\tau})A^* \tag{2.55}$$

where * denotes complex conjugation.

The proper complex Lorentz group is the direct product of SL(2, C) groups representing the proper Lorentz group: SL(2,C)⊗SL(2,C). The direct product is represented by the set of all pairs of 2 × 2 complex matrices of determinant one that satisfy the multiplication rule:

$$(A, B)(C, D) = (AC, BD) \tag{2.56}$$

To each pair of 2 × 2 matrices there corresponds a 4 × 4 complex matrix Λ(A, B). These complex matrices satisfy

$$\Lambda(A, B)\Lambda(C, D) = \Lambda(AC, BD) \tag{2.57}$$

The coordinate transformations of this complex Lorentz group representation has the form

$$x'^0 I + \mathbf{x}'\cdot\boldsymbol{\tau} = A(x^0 I + \mathbf{x}\cdot\boldsymbol{\tau})B^T \tag{2.58}$$

where the superscript T represents transpose. If we express A and B in terms of 2 × 2 matrices:

$$A = A^0 I + \mathbf{A}\cdot\boldsymbol{\tau} \quad \text{and} \quad B = B^0 I + \mathbf{B}\cdot\boldsymbol{\tau} \tag{2.59}$$

we must use complex values for the parameters A^0, **A**, B^0, and **B**.

2.7 A Different Complex Lorentz Group, L_C

There is another possible formulation of a complex homogeneous Lorentz group based on the complexification[45] of the 4 × 4 matrix proper, orthochronous representation of boosts, with matrices $\Lambda(\omega, \hat{\mathbf{u}})$, discussed earlier together with complex spatial rotations, the parity transformation and the time reversal transformation and products thereof. We will denote this group as homogeneous L_C since its matrices contain complex entries. L_C has 6 complex parameters – three complex boost parameters and three complex rotation parameters just as the standard formulation of the homogeneous complex Lorentz group SL(2,C)⊗SL(2,C). *L_C only has three boost operators and three rotation operators.*

It can be shown that the transformations of L_C satisfy the defining relation of the Lorentz group, eq. 2.1 since the algebraic proof of eq. 2.1 does not depend on the reality of the parameters. Thus the group L_C is a subgroup of the complex Lorentz group L(C).

In this subsection we will describe some of the basic features 4 × 4 matrix representation of L_C for use in the next chapter in the definition of a set of four spin ½ wave equations, and wave functions, that correspond to the four species of leptons and quarks of each generation.

The group elements of (homogeneous) L_C can be expressed as

$$\Lambda_C = \exp[i(\omega_r\hat{\mathbf{u}}_r + i\omega_i\hat{\mathbf{u}}_i)\cdot\mathbf{K} + i\boldsymbol{\theta}_c\cdot\mathbf{J}] \qquad (2.60)$$

where the vector $\boldsymbol{\theta}_c$ is a complex 3-vector, $\omega_r \geq 0$ and $\omega_i \geq 0$ are real numbers, and $\hat{\mathbf{u}}_r$ and $\hat{\mathbf{u}}_i$ are real normalized 3-vectors such that $\hat{\mathbf{u}}_r\cdot\hat{\mathbf{u}}_r = 1 = \hat{\mathbf{u}}_i\cdot\hat{\mathbf{u}}_i$. The generators of the homogeneous group are **K**, and **J** just as for the real Lorentz group.

Matrix Representation of L_C Group Boost Transformations

We shall now turn our attention to L_C boosts because they will be crucial in the determination of the equations of motion of various types of spin ½ particles in chapter 3. From this analysis we will be able to reestablish the previous results of this chapter

---

[45] Just as in the case of the SL(2,C)⊗SL(2,C) representation where the parameters of both A and B (eq. 2.59) are complex.

from a slightly deeper, more physical, viewpoint as well as to develop a more general picture.

An L_C boost can be expressed in the form

$$\Lambda_C(\mathbf{v_c}) = \exp[i\omega\hat{\mathbf{w}}\cdot\mathbf{K}] \tag{2.61}$$

where

$$\omega = (\omega_r^2 - \omega_i^2 + 2i\omega_r\omega_i\, \hat{\mathbf{u}}_r\cdot\hat{\mathbf{u}}_i)^{1/2} \tag{2.62}$$

and

$$\hat{\mathbf{w}} = (\omega_r\hat{\mathbf{u}}_r + i\omega_i\hat{\mathbf{u}}_i)/\omega \tag{2.63}$$

Since $\hat{\mathbf{u}}_r\cdot\hat{\mathbf{u}}_r = 1 = \hat{\mathbf{u}}_i\cdot\hat{\mathbf{u}}_i$

$$\hat{\mathbf{w}}\cdot\hat{\mathbf{w}} = 1 \tag{2.64a}$$

and the complex relative velocity is

$$\mathbf{v_c} = \hat{\mathbf{w}}\tanh(\omega) \tag{2.64b}$$

in analogy with eqs. 2.3 and 2.9.

Having placed L_C boost transformations in the form of eq. 2.3 we can take advantage of the form of real proper orthochronous Lorentz boost transformations, eq. 2.4, and analytically continue to complex ω and complex unit vectors $\hat{\mathbf{w}}$ provided eq. 2.64 is satisfied. The resulting complex generalization will be the matrix form of proper L_C boosts:

$$\Lambda_C(\mathbf{v_c}) = \exp[i\omega\hat{\mathbf{w}}\cdot\mathbf{K}] \equiv \Lambda_C(\omega, \hat{\mathbf{w}})$$

$$= \begin{bmatrix} \cosh(\omega) & -\sinh(\omega)\hat{w}_x & -\sinh(\omega)\hat{w}_y & -\sinh(\omega)\hat{w}_z \\ -\sinh(\omega)\hat{w}_x & 1 + (\cosh(\omega) - 1)\hat{w}_x^2 & (\cosh(\omega) - 1)\hat{w}_x\hat{w}_y & (\cosh(\omega) - 1)\hat{w}_x\hat{w}_z \\ -\sinh(\omega)\hat{w}_y & (\cosh(\omega) - 1)\hat{w}_x\hat{w}_y & 1 + (\cosh(\omega) - 1)\hat{w}_y^2 & (\cosh(\omega) - 1)\hat{w}_y\hat{w}_z \\ -\sinh(\omega)\hat{w}_z & (\cosh(\omega) - 1)\hat{w}_x\hat{w}_z & (\cosh(\omega) - 1)\hat{w}_y\hat{w}_z & 1 + (\cosh(\omega) - 1)\hat{w}_z^2 \end{bmatrix} \quad (2.65)$$

Since analytic continuations are unique, the above form for $\Lambda_C(\mathbf{v_c})$ is well-defined and unique. It spans the complete set of proper L_C boosts.

Connection to the Extended Lorentz Groups

The form of Left-handed and Right-handed Extended Lorentz group boosts, eqs. 2.8, 2.14 and 2.33, suggests we re-express eq. 2.57 in a different form to bring out the connection of L_C boosts with Left-handed and Right-handed Extended Lorentz group boosts.

Left-handed Extended Lorentz Group

If we let

$$\hat{\mathbf{u}}_\mathbf{i} = \hat{\mathbf{u}}_\mathbf{r} \equiv \hat{\mathbf{u}} \quad (2.66)$$

so that the vector $\hat{\mathbf{u}}_\mathbf{i}$ is parallel to $\hat{\mathbf{u}}_\mathbf{r}$, and

$$\omega_\mathbf{i} = \pi/2 \quad (2.67)$$

then $\Lambda_C(\mathbf{v_c})$ becomes a Left-handed Extended Lorentz boost:

$$\Lambda_C(\mathbf{v_c}) = \Lambda_L(\omega_\mathbf{r}, \mathbf{u}) \quad (2.68)$$

by eq. 2.28.

Right-handed Extended Lorentz Group

If we let

$$\hat{\mathbf{u}}_i = -\hat{\mathbf{u}}_r \equiv -\hat{\mathbf{u}} \tag{2.69}$$

so that the vector $\hat{\mathbf{u}}_i$ is anti-parallel to $\hat{\mathbf{u}}_r$, and

$$\omega_i = -\pi/2 \tag{2.70}$$

then $\Lambda_C(\mathbf{v}_c)$ becomes a Right-handed Extended Lorentz boost:

$$\Lambda_C(\mathbf{v}_c) = \Lambda_R(\omega_r\,,\, \mathbf{u}) \tag{2.71}$$

by eq. 2.33.

Difference between the Extended Lorentz Groups Reduced to Parallelism of $\hat{\mathbf{u}}_r$ and $\hat{\mathbf{u}}_i$

Since the Left-handed Extended Lorentz group leads to the Standard Model's left-handed features, it seems that the parallel case $\hat{\mathbf{u}}_i = \hat{\mathbf{u}}_r \equiv \hat{\mathbf{u}}$ is favored by Nature.[46] To some extent this concept of parallel vectors $\hat{\mathbf{u}}_i$ and $\hat{\mathbf{u}}_r$, which leads to the Left-handed Extended Lorentz group, is more intuitively satisfying then the anti-parallel case that leads to the Right-handed Extended Lorentz group. However, a deeper reason for Nature's choice remains to be found.

---

[46] It is possible that parity violation might disappear at ultra-high energies. Then we would view the parity symmetric theory as broken to the left-handed Standard Model currently established by experiment with right-handed parts at higher energy.

3. Free Spin ½ Particles – Leptons & Quarks

3.0 Chapter Overview

In this chapter we begin by developing dynamical equations for spin ½ particles based on the Extended Lorentz groups, and L_C, described in chapter 2. These spin ½ particles are conventional Dirac particles (Majorana particles are also allowed), spin ½ tachyons, and "color" versions of both types totaling four species. We will identify leptons and quarks with these fields.

3.1 Introduction

Tachyons are particles that move faster than the speed of light. As we saw in chapter 0 tachyons exist inside Black Holes, and within current theories – particularly SuperString theories.

Attempts to create canonical tachyon quantum field theories began in the 1960's. These attempts were made within the framework of the Lorentz group and, consequently, were limited to spin 0 theories since there are no finite dimensional representations of the Lorentz group for negative m^2 except for the one-dimensional representation. None of these attempts, or attempts since then, succeeded in creating a canonically quantized spin 0 tachyon quantum field theory.[47]

In this chapter we will formulate a free spin ½ tachyon Quantum Field theory. We choose to develop a spin ½ tachyon theory first because spin ½ particles (quarks and leptons) play an extraordinary role in the Standard Model. In chapter 4 we will consider bosonic tachyons.

We will develop our spin ½ tachyon theory from the "ground up" by applying a Left-Handed Extended Lorentz boost to the Dirac equation, and the Dirac spinor wave function, for a particle at rest. This procedure will give a tachyon spinor wave function, and the momentum space tachyon equation equivalent of the Dirac equation. Then we

---

[47] Except Blaha (2006), the first edition of this book.

will obtain the coordinate space tachyon Dirac equation, define a lagrangian, and proceed to create a canonical quantum field theory for spin ½ tachyons.

3.2 A Method for Deriving the Dirac Equation

In this section we will review a method of obtaining the Dirac equation by Lorentz boosts of the spinor wave function of a particle at rest. In the case of a Lorentz transformation the 4 x 4 matrix form of a Lorentz transformation of the Dirac matrices is

$$S^{-1}(\Lambda(v))\gamma^{\nu}S(\Lambda(v)) = \Lambda^{\nu}{}_{\mu}(v)\gamma^{\mu} \tag{3.1}$$

where $S(\Lambda(v))$ is

$$S(\Lambda(v)) = \exp(-i\omega\sigma_{0i}v_i/(2|\mathbf{v}|)) = \exp(-\omega\gamma^0\boldsymbol{\gamma}\cdot\mathbf{v}/(2|\mathbf{v}|))$$

$$= \cosh(\omega/2)I + \sinh(\omega/2)\gamma^0\boldsymbol{\gamma}\cdot\mathbf{p}/|\mathbf{p}| \tag{3.2}$$

with $\omega = \text{arctanh}(|\mathbf{v}|)$, $\cosh(\omega/2) = [(E+m)/(2m)]^{1/2}$ and $\sinh(\omega/2) = |\mathbf{p}|[2m(E+m)]^{-1/2}$. Also

$$S^{-1}(\Lambda(v)) = \gamma^0 S^{\dagger}(\Lambda(v))\gamma^0 = \exp(\omega\gamma^0\boldsymbol{\gamma}\cdot\mathbf{v}/(2|\mathbf{v}|))$$

$$= \cosh(\omega/2)I - \sinh(\omega/2)\gamma^0\boldsymbol{\gamma}\cdot\mathbf{p}/|\mathbf{p}| \tag{3.3}$$

A generic positive energy plane wave solution of the Dirac equation for a particle at rest with rest energy m is

$$\psi(x) = e^{-imt}w(0) \tag{3.4}$$

with w(0) a four component spinor column vector. It satisfies the momentum space Dirac equation for a particle at rest:

$$(m\gamma^0 - m)e^{-imt}w(0) = 0 \tag{3.5}$$

If we now apply $S(\Lambda(v))$ we find

$$0 = S(\Lambda(v))(m\gamma^0 - m)e^{-imt}w(0) = [mS(\Lambda(v))\gamma^0 S^{-1}(\Lambda(v)) - m]S(\Lambda(v))w(0)$$

A straightforward evaluation shows

$$mS(\Lambda(v))\gamma^0 S^{-1}(\Lambda(v)) = g_{\mu\nu}p^\mu\gamma^\nu = \not{p} \tag{3.6}$$

where $p^0 = (p^2 + m^2)^{1/2}$, $\mathbf{p} = \gamma m\mathbf{v}$, and $p = |\mathbf{p}|$. In addition

$$S(\Lambda(v))w(0) = w(p) \tag{3.7}$$

is a positive energy Dirac spinor. Therefore the Dirac equation in momentum space has the form:

$$(\not{p} - m)e^{-ip\cdot x}w(p) = 0 \tag{3.8}$$

where the exponential factor, mt, is also boosted to $p\cdot x$. Eq. 3.8 implies the free, coordinate space Dirac equation:

$$(i\gamma^\mu\partial/\partial x^\mu - m)\psi(x) = 0 \tag{3.9}$$

3.3 Derivation of the Tachyonic Dirac Equation

The Left-handed Extended Lorentz boost has the form:

$$\Lambda_L(\omega, \mathbf{u}) = \Lambda(\omega + i\pi/2, \mathbf{u}) = \exp[i\omega_L\hat{\mathbf{u}}\cdot\mathbf{K}] \tag{3.10}$$

where $\omega_L = \omega + i\pi/2$ and

$$\begin{aligned} \cosh(\omega_L) &= i\sinh(\omega) = -\gamma = i\gamma_s \\ \sinh(\omega_L) &= i\cosh(\omega) = -\beta\gamma = i\beta\gamma_s \end{aligned} \tag{2.11}$$

with, $\beta = v > 1$, $\gamma_s = (\beta^2 - 1)^{-1/2}$, and $\boldsymbol{\omega} \geq \mathbf{0}$. Thus

$$\sinh(\omega) = \gamma_s$$
$$\cosh(\omega) = \beta\gamma_s \tag{2.12}$$

The corresponding spinor transformation is:

$$S_L(\Lambda_L(\omega, \mathbf{u})) = \exp(-i\omega_L\sigma_{0i}v_i/(2|\mathbf{v}|)) = \exp(-\omega_L\gamma^0\boldsymbol{\gamma}\cdot\mathbf{v}/(2|\mathbf{v}|))$$
$$= \cosh(\omega_L/2)I + \sinh(\omega_L/2)\gamma^0\boldsymbol{\gamma}\cdot\mathbf{p}/|\mathbf{p}| \tag{3.11}$$

The inverse transformation is

$$S_L^{-1}(\Lambda_L(\omega, \mathbf{u})) = \gamma^2\gamma^0K^{-1}S_L{}^{\dagger}K\gamma^0\gamma^2 = \gamma^2\gamma^0S_L{}^{T}\gamma^0\gamma^2 = \exp(\omega_L\gamma^0\boldsymbol{\gamma}\cdot\mathbf{v}/(2|\mathbf{v}|))$$
$$= \cosh(\omega_L/2)I - \sinh(\omega_L/2)\gamma^0\boldsymbol{\gamma}\cdot\mathbf{p}/|\mathbf{p}| \tag{3.12}$$

where the superscript T denotes the transpose and K is the complex conjugation operator (that also appears in the time-reversal operator). Note that S_L is not unitary just as the equivalent spinor Lorentz transformation $S(\Lambda(v))$ is not unitary.
We can now apply a left-handed superluminal transformation to the generic positive energy plane wave solution of the Dirac equation for a particle of mass m at rest. The result is

$$0 = S_L(\Lambda_L(\omega, \mathbf{u}))(m\gamma^0 - m)e^{-imt}w(0)$$
$$= [mS_L\gamma^0S_L{}^{-1} - m]e^{-imt}S_Lw(0)$$

where $S_L = S_L(\Lambda_L(\omega, \mathbf{u}))$. After some algebra

$$mS_L\gamma^0S_L{}^{-1} = m[\cosh(\omega_L)\gamma^0 - \sinh(\omega_L)\boldsymbol{\gamma}\cdot\mathbf{p}/|\mathbf{p}|]$$

$$= i\gamma^0 E - i\boldsymbol{\gamma}\cdot\mathbf{p} = i\not{p} \qquad (3.13)$$

using eqs. 2.11 and the tachyon energy and momentum expressions

$$\mathbf{p} = m\mathbf{v}\gamma_s \qquad\qquad E = m\gamma_s \qquad (3.14)$$

Also

$$S_L w(0) = w_T(p') \qquad (3.15)$$

is a tachyon spinor. See Appendix 3-A (at the end of this chapter) for a discussion of tachyon spinors.

The momentum space tachyonic Dirac equation is

$$(i\not{p} - m)e^{ip\cdot x}w_T(p) = 0 \qquad (3.16)$$

where $p\cdot x = Et - \mathbf{p}\cdot\mathbf{x}$ after performing a corresponding left-handed superluminal coordinate transformation in the exponential factor based on eq. 2.10d. Thus the positive energy wave is transformed into a negative energy wave by the superluminal transformation.

If we apply $i\not{p}$ to we find the tachyon mass condition is satisfied

$$-E^2 + \mathbf{p}^2 = m^2 \qquad (3.17)$$

Transforming back to coordinate space we obtain the *tachyonic Dirac equation*:

$$(\gamma^\mu\partial/\partial x^\mu - m)\psi_T(x) = 0 \qquad (3.18)$$

The "missing" factor of i in the first term of eq. 3.18 requires the lagrangian to be different from the conventional Dirac lagrangian in order for the lagrangian to be real. The simplest, physically acceptable, free spin ½ tachyon lagrangian density is:

$$\mathcal{L}_T = \psi_T^S(\gamma^\mu\partial/\partial x^\mu - m)\psi_T(x) \tag{3.19}$$

where

$$\psi_T^S = \psi_T^\dagger i\gamma^0\gamma^5 \tag{3.20}$$

The corresponding action is

$$I = \int d^4x \mathcal{L}_T \tag{3.21}$$

Appendix 3-B proves I is real. The Hamiltonian density is

$$\mathcal{H} = \pi_T\dot{\psi}_T - \mathcal{L} = i\psi_T^\dagger\gamma^5(\boldsymbol{\alpha}\cdot\nabla + \beta m)\psi_T = -i\psi_T^\dagger\gamma^5\dot{\psi}_T \tag{3.22}$$

using the tachyon Dirac equation to obtain the last equality. The reader will note that the tachyon hamiltonian is hermitean by explicit calculation up to an irrelevant total spatial divergence.

Probability Conservation Law

The tachyon Dirac equation implies a probability conservation law:

$$\partial\rho_5/\partial t = \nabla\cdot\mathbf{j}_5 \tag{3.23}$$

where

$$\rho_5 = \psi_T^\dagger\gamma^5\psi_T \qquad\qquad \mathbf{j}_5 = \psi_T^\dagger\gamma^5\boldsymbol{\alpha}\psi_T \tag{3.24}$$

We are thus led to define the conserved axial charge Q_5

$$Q_5 = \int d^3x\, \psi_T^\dagger\gamma^5\psi_T \tag{3.25}$$

Energy-Momentum Tensor

The tachyon energy-momentum tensor is

$$\mathfrak{T}_{T\mu\nu} = -\, g_{\mu\nu}\, \mathcal{L}_T + \partial\mathcal{L}_T/\partial(\partial\psi_T/\partial x_\mu)\, \partial\psi_T/\partial x^\nu \tag{3.26}$$

$$= i\psi_T^{\dagger}\gamma^0\gamma^5\gamma_\mu\partial\psi_T/\partial x^\nu \tag{3.27}$$

and thus the conserved energy and momentum are

$$P^0 = H = \int d^3x\, \mathfrak{T}_T^{00} = i\int d^3x\psi_T^{\dagger}\gamma^5(\boldsymbol{\alpha}\cdot\nabla + \beta m)\psi_T \tag{3.28}$$

and

$$P^I = \int D^3X\, \mathfrak{T}_T^{0I} = -\, I\int D^3X\, \Psi_T^{\dagger}\Gamma^5\partial\Psi_T/\partial X_I \tag{3.29}$$

Both the energy and momentum differ significantly from the corresponding quantities for conventional Dirac fields.

3.4 Tachyon Canonical Quantization

Having defined a suitable tachyon lagrangian we can now proceed to its canonical quantization. The conjugate momentum can be calculated from the lagrangian density eq. 3.19:

$$\pi_{Ta} = \partial\, \mathcal{L}_T/\partial\dot{\psi}_{Ta} \equiv \partial\mathcal{L}_T/\partial(\partial\psi_{Ta}/\partial t) = -i(\psi_T^{\dagger}\gamma^5)_a \tag{3.30}$$

The resulting non-zero, canonical anti-commutation relations are

$$\{\pi_{Ta}(x), \psi_{Tb}(x')\} = i\, \delta_{ab}\, \delta^3(x - x')$$

or

$$\{\psi_{T\,a}^{\dagger}(x), \psi_{Tb}(x')\} = -\, [\gamma^5]_{ab}\, \delta^3(x - x') \tag{3.31}$$

At this point we might attempt to complete the canonical quantization procedure in the conventional manner by fourier expanding the field and specifying anti-commutation relations for the fourier component amplitudes. However the incompleteness of the set of plane waves, which are limited by the restriction $|\mathbf{p}| \geq m$, causes the anti-commutator of the fields *not* to yield a $\delta^3(x - x')$. Thus the conventional approach fails to yield the required anti-commutation relations.[48]

Other approaches: 1) decompose the tachyon field into left-handed and right-handed parts and then second quantize each part; and 2) second quantize in light-front coordinates ($x^{\pm} = (x^0 \pm x^3)/\sqrt{2}$). These approaches also both fail.[49]

The only approach that does succeed[50] is to decompose the tachyon field into left-handed and right-handed parts and then second quantize in light-front coordinates. We follow that procedure in the following subsections.

Separation into Left-Handed and Right-Handed Fields

We will use a transformed set of Dirac matrices to develop our left-handed and right-handed tachyon formulations:

$$\gamma^0 = \begin{bmatrix} 0 & -I \\ -I & 0 \end{bmatrix} \qquad \gamma^i = \begin{bmatrix} 0 & \sigma_i \\ -\sigma_i & 0 \end{bmatrix} \qquad \gamma^5 = \begin{bmatrix} I & 0 \\ 0 & -I \end{bmatrix} \tag{3.32}$$

which are obtained from the usual Dirac matrices by applying the unitary transformation $U = 2^{-\frac{1}{2}}(I + \gamma^5\gamma^0)$. I is the 4×4 identity matrix. The γ^5 chirality operator's eigenvalues define handedness: +1 corresponds to right-handed; and −1 corresponds to left-handed:

$$\gamma^5\psi_L = -\psi_L \qquad\qquad \gamma^5\psi_R = \psi_R \tag{3.33}$$

---

[48] See G. Feinberg, Phys. Rev. **159**, 1089 (1967) for example.
[49] See the first edition Blaha (2006) where these possibilities were considered and found to fail.
[50] Blaha (2006) discusses this case in detail.

Consequently, we can define left-handed and right-handed tachyon fields with the projection operators:

$$\begin{aligned} C^{\pm} &= \tfrac{1}{2}(I \pm \gamma^5) \\ C^{+} + C^{-} &= I \\ C^{\pm\,2} &= C^{\pm} \\ C^{+}C^{-} &= 0 \end{aligned} \qquad (3.34)$$

with the result

$$\begin{aligned} \psi_{TL} &= C^{-}\psi_{T} \\ \psi_{TR} &= C^{+}\psi_{T} \end{aligned} \qquad (3.35)$$

We can calculate the commutation relations of the left-handed and right-handed tachyon fields from eq. 3.31 by pre-multiplying and post-multiplying by $\tfrac{1}{2}(1 - \gamma^5)$ and $\tfrac{1}{2}(1 + \gamma^5)$. The results are:

$$\{\psi_{TLa}^{\dagger}(x), \psi_{TLb}(x')\} = \tfrac{1}{2}(1 - \gamma^5)_{ab}\, \delta^3(x - x') \qquad (3.36)$$

$$\{\psi_{TRa}^{\dagger}(x), \psi_{TRb}(x')\} = -\tfrac{1}{2}(1 + \gamma^5)_{ab}\, \delta^3(x - x') \qquad (3.37)$$

$$\{\psi_{TLa}^{\dagger}(x), \psi_{TRb}(x')\} = \{\psi_{TRa}^{\dagger}(x), \psi_{TLb}(x')\} = 0 \qquad (3.38)$$

The lagrangian density of eq. 3.19 decomposes into left-handed and right-handed parts:

$$\mathcal{L}_T = \psi_{TL}^{\dagger}\gamma^0 i\gamma^{\mu}\partial_{\mu}\psi_{TL} - \psi_{TR}^{\dagger}\gamma^0 i\gamma^{\mu}\partial_{\mu}\psi_{TR} - im[\psi_{TR}^{\dagger}\gamma^0\psi_{TL} - \psi_{TL}^{\dagger}\gamma^0\psi_{TR}] \qquad (3.39)$$

Further Separation into + and – Light-Front Fields

There have been many studies of light-front (infinite momentum frame) physics in the past forty years.[51] Light-front coordinates cannot be obtained by a Lorentz transformation, or by a superluminal transformation, from a standard set of coordinate system variables even in a limiting sense. Instead they are a defined set of variables that have been used to develop quantum field theories that have been shown to be equivalent to quantum field theories based on conventional coordinates. In particular, light-front quantum field theories have been shown to yield fully Lorentz covariant S matrix elements that are the same as S matrix elements calculated in the conventional way.

Light-front variables can be defined by:

$$x^{\pm} = (x^0 \pm x^3)/\sqrt{2}$$
$$\partial/\partial x^{\pm} \equiv \partial^{\mp} \equiv (\partial/\partial x^0 \pm \partial/\partial x^3)/\sqrt{2} \tag{3.40}$$

with the "transverse" coordinate variables, x^1 and x^2, unchanged.

The inner product of two 4-vectors has the form

$$x{\cdot}y = x^+y^- + y^+x^- - x^1y^1 - x^2y^2 \tag{3.41}$$

and the light-front definition of Dirac matrices is:

$$\gamma^{\pm} = (\gamma^0 \pm \gamma^3)/\sqrt{2} \tag{3.42}$$

with transverse matrices γ^1 and γ^2 defined as usual. Note the useful identity:

---

[51] L. Susskind, Phys. Rev. **165**, 1535 (1968); K. Bardakci and M. B. Halpern Phys. Rev. **176**, 1686 (1968), S. Weinberg, Phys. Rev. **150**, 1313 (1966); J. Kogut and D. Soper, Phys. Rev. **D1**, 2901 (1970); J. D. Bjorken, J. Kogut, and D. Soper, Phys. Rev. **D3**, 1382 (1971); R. A. Neville and F. Rohrlich, Nuov. Cim. **A1**, 625 (1971); F. Rohrlich, Acta Phys Austr. Suppl. **8**, 277 (1971); S-J Chang, R. Root, and T-M Yan, Phys. Rev. **D7**, 1133 (1973); S-J Chang, and T-M Yan, Phys. Rev. **D7**, 1147 (1973); T-M Yan, Phys. Rev. **D7**, 1761 (1973); T-M Yan, Phys. Rev. **D7**, 1780 (1973); C. Thorn, Phys. Rev. **D19**, 639 (1979); and references therein.

$$\gamma^{\pm\,2} = 0$$

We define "+" and "–" tachyon fields with the projection operators:

$$R^{\pm} = ½(I \pm \gamma^0\gamma^3) \qquad (3.43)$$

Left-handed, ± light-front fields: $\psi_{TL}{}^{\pm} = R^{\pm}C^{-}\psi_T$

(3.44)

Right-handed, ± light-front fields: $\psi_{TR}{}^{\pm} = R^{\pm}C^{+}\psi_T$

Now if we transform to light-front variables and fields as above we obtain the light-front free tachyon lagrangian:

$$\begin{aligned}\mathcal{L}_T = {} & 2^{½}\psi_{TL}{}^{+\dagger}i\partial^{-}\psi_{TL}{}^{+} + 2^{½}\psi_{TL}{}^{-\dagger}i\partial^{+}\psi_{TL}{}^{-} - \psi_{TL}{}^{+\dagger}\gamma^0 i\gamma^j\partial^j\psi_{TL}{}^{-} - \psi_{TL}{}^{-\dagger}\gamma^0 i\gamma^j\partial^j\psi_{TL}{}^{+} - \\ & - 2^{½}\psi_{TR}{}^{+\dagger}i\partial^{-}\psi_{TR}{}^{+} - 2^{½}\psi_{TR}{}^{-\dagger}i\partial^{+}\psi_{TR}{}^{-} + \psi_{TR}{}^{+\dagger}\gamma^0 i\gamma^j\partial^j\psi_{TR}{}^{-} + \psi_{TR}{}^{-\dagger}\gamma^0 i\gamma^j\partial^j\psi_{TR}{}^{+} - \\ & - im[\psi_{TR}{}^{+\dagger}\gamma^0\psi_{TL}{}^{-} - \psi_{TL}{}^{+\dagger}\gamma^0\psi_{TR}{}^{-} + \psi_{TR}{}^{-\dagger}\gamma^0\psi_{TL}{}^{+} - \psi_{TL}{}^{-\dagger}\gamma^0\psi_{TR}{}^{+}]\end{aligned} \qquad (3.45)$$

with implied sums over j = 1,2. In contrast to the light-front tachyon lagrangian we note the corresponding light-front Dirac fermion lagrangian is

$$\begin{aligned}\mathcal{L}_D = {} & 2^{½}\psi_{L}{}^{+\dagger}i\partial^{-}\psi_{L}{}^{+} + 2^{½}\psi_{L}{}^{-\dagger}i\partial^{+}\psi_{L}{}^{-} - \psi_{L}{}^{+\dagger}\gamma^0 i\gamma^j\partial^j\psi_{L}{}^{-} - \psi_{L}{}^{-\dagger}\gamma^0 i\gamma^j\partial^j\psi_{L}{}^{+} - \\ & + 2^{½}\psi_{R}{}^{+\dagger}i\partial^{-}\psi_{R}{}^{+} + 2^{½}\psi_{R}{}^{-\dagger}i\partial^{+}\psi_{R}{}^{-} - \psi_{R}{}^{+\dagger}\gamma^0 i\gamma^j\partial^j\psi_{R}{}^{-} - \psi_{R}{}^{-\dagger}\gamma^0 i\gamma^j\partial^j\psi_{R}{}^{+} - \\ & - im[\psi_{R}{}^{+\dagger}\gamma^0\psi_{L}{}^{-} + \psi_{L}{}^{+\dagger}\gamma^0\psi_{R}{}^{-} + \psi_{R}{}^{-\dagger}\gamma^0\psi_{L}{}^{+} + \psi_{L}{}^{-\dagger}\gamma^0\psi_{R}{}^{+}]\end{aligned} \qquad (3.46)$$

The difference in signs between eqs. 3.45 and 3.46 will turn out to be a crucial factor in the derivation of features of the Standard Model later.

Returning to the tachyon lagrangian eq. 3.45 we obtain equations of motion through the standard variational techniques:

$$\begin{aligned}
2^{1/2}i\partial^{-}\psi_{TL}^{\;+} - \gamma^0 i\gamma^j\partial^j\psi_{TL}^{\;-} + im\gamma^0\psi_{TR}^{\;-} &= 0 \\
2^{1/2}i\partial^{-}\psi_{TR}^{\;+} - \gamma^0 i\gamma^j\partial^j\psi_{TR}^{\;-} + im\gamma^0\psi_{TL}^{\;-} &= 0 \\
2^{1/2}i\partial^{+}\psi_{TL}^{\;-} - \gamma^0 i\gamma^j\partial^j\psi_{TL}^{\;+} + im\gamma^0\psi_{TR}^{\;+} &= 0 \\
2^{1/2}i\partial^{+}\psi_{TR}^{\;-} - \gamma^0 i\gamma^j\partial^j\psi_{TR}^{\;+} + im\gamma^0\psi_{TL}^{\;+} &= 0
\end{aligned} \tag{3.47}$$

Eqs. 3.47 show that $\psi_{TL}^{\;-}$ and $\psi_{TR}^{\;-}$ are dependent fields that are functions of $\psi_{TL}^{\;+}$ and $\psi_{TR}^{\;+}$ on the light-front where x^+ equals a constant. They can be expressed in an integral form as well. (The independent fields $\psi_{TL}^{\;+}$ and $\psi_{TR}^{\;+}$ play a fundamental role in tachyon theory and are used to define "in" and "out" tachyon states in perturbation theory.)

The conjugate momenta implied by eq. 3.45 are

$$\pi_{TL}^{\;+} = \partial\mathfrak{L}/\partial(\partial^{-}\psi_{TL}^{\;+}) = 2^{1/2}i\psi_{TL}^{\;+\dagger} \tag{3.48}$$
$$\pi_{TL}^{\;-} = \partial\mathfrak{L}/\partial(\partial^{-}\psi_{TL}^{\;-}) = 0$$
$$\pi_{TR}^{\;+} = \partial\mathfrak{L}/\partial(\partial^{-}\psi_{TR}^{\;+}) = -2^{1/2}i\psi_{TR}^{\;+\dagger} \tag{3.49}$$
$$\pi_{TR}^{\;-} = \partial\mathfrak{L}/\partial(\partial^{-}\psi_{TR}^{\;-}) = 0$$

Quantization on surfaces of constant x^+ (light-front surfaces) has been shown to support satisfactory formulations of Quantum Electrodynamics and other quantum field theories. Thus x^+ plays the role of the "time" variable in light-front quantized theories. So we will define canonical equal x^+ anti-commutation relations for spin ½ tachyons.

The resulting canonical equal-light-front ($x^+ = y^+$) anti-commutation relations of the independent fields are:

$$\{\psi_{TL}^{\;+\dagger}{}_a(x), \psi_{TL}^{\;+}{}_b(y)\} = 2^{-1}[C^{-}R^{+}]_{ab}\,\delta(x^{-} - y^{-})\delta^2(x - y) \tag{3.50}$$

$$\{\psi_{TR}{}^{+\dagger}{}_{a}(x), \psi_{TR}{}^{+}{}_{b}(y)\} = -2^{-1}[C^{+}R^{+}]_{ab}\,\delta(x^{-} - y^{-})\delta^{2}(x - y) \qquad (3.51)$$

$$\{\psi_{TL}{}^{+}{}_{a}{}^{\dagger}(x), \psi_{TR}{}^{+}{}_{b}(y)\} = \{\psi_{TR}{}^{+}{}_{a}{}^{\dagger}(x), \psi_{TL}{}^{+}{}_{b}(y)\} = 0 \qquad (3.52)$$

$$\{\psi_{TL}{}^{+}{}_{a}(x), \psi_{TR}{}^{+}{}_{b}(y)\} = \{\psi_{TR}{}^{+}{}_{a}{}^{\dagger}(x), \psi_{TL}{}^{+\dagger}{}_{b}(y)\} = 0 \qquad (3.53)$$

where the factors of 2^{-1} are the result of the $2^{1/2}$ factor in eqs. 3.48 and 3.49, and the factor of $2^{-1/2}$ in the definition of x^{-} in eq. 3.40.

If we compare eqs. 3.50 and 3.51 with the corresponding anti-commutation relations of conventional Dirac quantum fields:

$$\{\psi_{L}{}^{+\dagger}{}_{a}(x), \psi_{L}{}^{+}{}_{b}(y)\} = 2^{-1}[C^{-}R^{+}]_{ab}\,\delta(x^{-} - y^{-})\delta^{2}(x - y) \qquad (3.54)$$

$$\{\psi_{R}{}^{+\dagger}{}_{a}(x), \psi_{R}{}^{+}{}_{b}(y)\} = 2^{-1}[C^{+}R^{+}]_{ab}\,\delta(x^{-} - y^{-})\delta^{2}(x - y) \qquad (3.55)$$

we see that the right-handed tachyon anti-commutation relation (eq. 3.51) has a minus sign relative to the corresponding right-handed conventional anti-commutation relation (eq. 3.55). The right-handed tachyon anti-commutation relation (eq. 3.51) with its minus sign will require compensating minus signs in its creation and annihilation Fourier component operators' anti-commutation relations.

The sign differences between the lagrangian terms in eqs. 3.47 and 3.48 ultimately lead to parity violating features in the Standard Model lagrangian and thus resolve the long-standing question: Why parity violation? Answer: Nature chooses the Left-handed Extended Lorentz group. Thus the source of parity violation, and much of the form of the Standard Model, is in superluminal physics.

Left-Handed Tachyons

The free, "+" light-front, left-handed tachyon wave function Fourier expansion is:

$$\psi_{TL}^{+}(x) = \sum_{\pm s} \int d^2p dp^{+} N_{TL}^{+}(p)\theta(p^{+})[b_{TL}^{+}(p, s)u_{TL}^{+}(p, s)e^{-ip.x} + + d_{TL}^{+\dagger}(p, s)v_{TL}^{+}(p, s)e^{+ip.x}] \quad (3.56)$$

and its hermitean conjugate is

$$\psi_{TL}^{+\dagger}(x) = \sum_{\pm s} \int d^2p dp^{+} N_{TL}^{+}(p)\theta(p^{+}) [b_{TL}^{+\dagger}(p, s)u_{TL}^{+\dagger}(p,s)e^{+ip.x} + + d_{TL}^{+}(p, s)v_{TL}^{+\dagger}(p, s)e^{-ip.x}] \quad (3.57)$$

where † indicates hermitean conjugate, where

$$N_{TL}^{+}(p) = [2m|\mathbf{p}|/((2\pi)^3(p^{+}(p^{+} - p^{-}) + p_{\perp}^{2}))]^{1/2} \quad (3.57a)$$

where the anti-commutation relations of the Fourier coefficient operators are

$$\begin{aligned}
\{b_{TL}^{+}(q,s), b_{TL}^{+\dagger}(p,s')\} &= \delta_{ss'}\delta^2(\mathbf{q} - \mathbf{p})\delta(q^{+} - p^{+}) \\
\{d_{TL}^{+}(q,s), d_{TL}^{+\dagger}(p,s')\} &= \delta_{ss'}\delta^2(\mathbf{q} - \mathbf{p})\delta(q^{+} - p^{+}) \\
\{b_{TL}^{+}(q,s), b_{TL}^{+}(p,s')\} &= \{d_{TL}^{+}(q,s), d_{TL}^{+}(p,s')\} = 0 \\
\{b_{TL}^{+\dagger}(q,s), b_{TL}^{+\dagger}(p,s')\} &= \{d_{TL}^{+\dagger}(q,s), d_{TL}^{+\dagger}(p,s')\} = 0 \\
\{b_{TL}^{+}(q,s), d_{TL}^{+\dagger}(p,s')\} &= \{d_{TL}^{+}(q,s), b_{TL}^{+\dagger}(p,s')\} = 0 \\
\{b_{TL}^{+\dagger}(q,s), d_{TL}^{+\dagger}(p,s')\} &= \{d_{TL}^{+}(q,s), b_{TL}^{+}(p,s')\} = 0
\end{aligned} \quad (3.58)$$

and where the spinors are

$$u_{TL}^{+}(p, s) = C^{-} R^{+} S_L(\Lambda_L(\mathbf{p}))w^1(0)$$

$$u_{TL}^{\;+}(p, -s) = C^- R^+ S_L(\Lambda_L(\mathbf{p}))w^2(0)$$
$$v_{TL}^{\;+}(p, s) = C^- R^+ S_L(\Lambda_L(\mathbf{p}))w^3(0)$$
$$v_{TL}^{\;+}(p, -s) = C^- R^+ S_L(\Lambda_L(\mathbf{p}))w^4(0)$$

(3.59)

$$u_{TL}^{\;+\dagger}(p, s) = w^{1T}(0)S_L^{\dagger}(\Lambda_L(\mathbf{p}))R^+C^-$$
$$u_{TL}^{\;+\dagger}(p, -s) = w^{2T}(0)S_L^{\dagger}(\Lambda_L(\mathbf{p}))R^+C^-$$
$$v_{TL}^{\;+\dagger}(p, s) = w^{3T}(0)S_L^{\dagger}(\Lambda_L(\mathbf{p}))R^+C^-$$
$$v_{TL}^{\;+\dagger}(p, -s) = w^{4T}(0)S_L^{\dagger}(\Lambda_L(\mathbf{p}))R^+C^-$$

where the superscript "T" indicates the transpose. (These spinors are described in Appendix 3-A.)

The canonical left-handed, light-front anti-commutation relation (eq. 3.50) follows from eqs. 3.56 – 3.59:

$$\{\psi_{TL}^{\;+}{}_a(x), \psi_{TL}^{\;+\dagger}{}_b(y)\} = \sum_{\pm s,s'} \int d^2p dp^+ \int d^2p' dp'^+ \, N_{TL}^{\;+}(p) N_{TL}^{\;+}(p')\theta(p^+)\theta(p'^+)\cdot$$
$$\cdot[\{b_{TL}^{\;+\dagger}(p',s'), b_{TL}^{\;+}(p,s)\} u_{TL}^{\;+}{}_a(p,s) u_{TL}^{\;+\dagger}{}_b(p',s') e^{+ip'.y - ip.x} +$$
$$+ \{d_{TL}^{\;+}(p',s'), d_{TL}^{\;+\dagger}(p,s)\} v_{TL}^{\;+}{}_a(p,s) v_{TL}^{\;+\dagger}{}_b(p',s') e^{-ip'.y + ip.x}]$$

$$= \sum_{\pm s} \int d^2p dp^+ \, N_{TL}^{\;+2}(p)\theta(p^+)[u_{TL}^{\;+}{}_a(p,s) u_{TL}^{\;+\dagger}{}_b(p,s) e^{+ip.(y-x)} +$$
$$+ v_{TL}^{\;+}{}_a(p,s) v_{TL}^{\;+\dagger}{}_b(p,s) e^{-ip.(y-x)}]$$

$$= -i\int d^2p dp^+ \theta(p^+) N_{TL}^{+2}(p)(2m|\mathbf{p}|)^{-1}\{[C^-R^+(i\not{p} - m)\gamma\cdot\mathbf{p}R^+C^-]_{ab}e^{+ip\cdot(y-x)} + + [C^-R^+(i\not{p} + m)\gamma\cdot\mathbf{p}R^+C^-]_{ab}e^{-ip\cdot(y-x)}\}$$

$$= -i\int d^2p_\perp \int_0^\infty dp^+ N_{TL}^{+2}(p)\{[C^-R^+(ip^+(p^+ - p^-) + ip_\perp{}^2 - mp_\perp\cdot\gamma_\perp)C^-]_{ab}\cdot \cdot e^{+ip^+(y^- - x^-) - ip_\perp\cdot(y_\perp - x_\perp)} - - [C^-R^+(-ip^+(p^+ - p^-) - ip_\perp{}^2 - mp_\perp\cdot\gamma_\perp)C^-]_{ab}e^{-ip^+(y^- - x^-) + ip_\perp\cdot(y_\perp - x_\perp)}\}/(2m|\mathbf{p}|)$$

$$= \int d^2p_\perp \int_{-\infty}^\infty dp^+ N_{TL}^{+2}(p)[C^-R^+(p^+(p^+ - p^-) + p_\perp{}^2)]_{ab}\cdot \cdot e^{+ip^+(y^- - x^-) - ip_\perp\cdot(y_\perp - x_\perp)}/(2m|\mathbf{p}|)$$

upon letting $p^+ \to -p^+$ and $\mathbf{p}_\perp \to -\mathbf{p}_\perp$ in the second term after using $N_{TL}^{+2}(p)(p^+(p^+ - p^-) + p_\perp{}^2) = 1$. The result

$$= \tfrac{1}{2}\int d^2p_\perp \int_{-\infty}^\infty dp^+ (2\pi)^{-3}[C^-R^+]_{ab}e^{+ip^+(y^- - x^-) - ip_\perp\cdot(y_\perp - x_\perp)}$$

$$= 2^{-1}[C^-R^+]_{ab}\delta(y^- - x^-)\delta^2(\mathbf{y} - \mathbf{x}) \qquad (3.60)$$

Therefore we have left-handed, light-front quantized tachyons with canonical commutation relations and localized tachyons. As a result we have a canonical tachyon Quantum Field Theory.

Right-Handed Tachyons

The case of right-handed tachyons is similar to the left-handed case with only two differences: a minus sign in the creation and annihilation operator anti-commutation relations, and the use of right-handed projection operators. The right-handed tachyon wave function light-front Fourier expansion is:

$$\psi_{TR}^{+}(x) = \sum_{\pm s} \int d^2p dp^{+} N_{TR}^{+}(p)\theta(p^{+})[b_{TR}^{+}(p, s)u_{TR}^{+}(p, s)e^{-ip\cdot x} + \\ + d_{TR}^{+\dagger}(p, s)v_{TR}^{+}(p, s)e^{+ip\cdot x}] \quad (3.61)$$

and its hermitean conjugate is

$$\psi_{TR}^{+\dagger}(x) = \sum_{\pm s} \int d^2p dp^{+} N_{TR}^{+}(p)\theta(p^{+}) [b_{TR}^{+\dagger}(p, s)u_{TR}^{+\dagger}(p, s)e^{+ip\cdot x} + \\ + d_{TR}^{+}(p, s)v_{TR}^{+\dagger}(p, s)e^{-ip\cdot x}] \quad (3.62)$$

where $N_{TR}^{+}(p) = N_{TL}^{+}(p)$, where the anti-commutation relations of the Fourier coefficient operators are

$$\{b_{TR}^{+}(q,s), b_{TR}^{+\dagger}(p,s')\} = -\delta_{ss'}\delta^2(\mathbf{q} - \mathbf{p})\delta(q^{+} - p^{+}) \quad (3.63)$$
$$\{d_{TR}^{+}(q,s), d_{TR}^{+\dagger}(p,s')\} = -\delta_{ss'}\delta^2(\mathbf{q} - \mathbf{p})\delta(q^{+} - p^{+})$$
$$\{b_{TR}^{+}(q,s), b_{TR}^{+}(p,s')\} = \{d_{TR}^{+}(q,s), d_{TR}^{+}(p,s')\} = 0$$
$$\{b_{TR}^{+\dagger}(q,s), b_{TR}^{+\dagger}(p,s')\} = \{d_{TR}^{+\dagger}(q,s), d_{TR}^{+\dagger}(p,s')\} = 0$$
$$\{b_{TR}^{+}(q,s), d_{TR}^{+\dagger}(p,s')\} = \{d_{TR}^{+}(q,s), b_{TR}^{+\dagger}(p,s')\} = 0$$
$$\{b_{TR}^{+\dagger}(q,s), d_{TR}^{+\dagger}(p,s')\} = \{d_{TR}^{+}(q,s), b_{TR}^{+}(p,s')\} = 0$$

and where the spinors are

$$u_{TR}^{+}(p, s) = C^{+}R^{+}u_T(p,s) \quad (3.64)$$

$$v_{TR}^{\;+}(p, s) = C^{+}R^{+}v_{T}(p,s) \tag{3.65}$$

by Appendix 3-A (eq. 3-A.7).

The right-handed anti-commutation relation (eq. 3.51) with the minus sign follows in particular because of the minus signs in eqs. 3.63.

Interpretation of Tachyon Creation and Annihilation Operators

To properly discuss the physical interpretation of tachyon creation and annihilation operators we must first determine the hamiltonian and momentum operators in terms of creation and annihilation operators.

The energy-momentum tensor density is the symmetrized version of

$$\mathfrak{T}^{\mu\nu} = \sum_i \partial\mathfrak{L}/\partial(\partial\chi_i/\partial x_\mu)\, \partial\chi_i/\partial x_\nu - g^{\mu\nu}\mathfrak{L} \tag{3.66}$$

where the sum over i is over the fields. The light-front hamiltonian is

$$H \equiv P^{-} = T^{+-} = \int dx^{-}d^{2}x\mathfrak{T}^{+-} \tag{3.67}$$

and the "momenta" are

$$P^{+} = T^{++} = \int dx^{-}d^{2}x\mathfrak{T}^{++} \tag{3.68}$$

$$P^{i} = T^{+i} = \int dx^{-}d^{2}x\mathfrak{T}^{+i} \tag{3.69}$$

for i = 1,2.

The light-front, left-handed and right-handed tachyon lagrangian $\mathfrak{L}_T$ is eq. 3.45 and its equations of motion are eqs. 3.47. They imply

$$H = i2^{-1/2}\int dx^{-}d^{2}x\, [\psi_{TL}^{\;+\dagger}\partial^{-}\psi_{TL}^{\;+} - \partial^{-}\psi_{TL}^{\;+\dagger}\psi_{TL}^{\;+} + \psi_{TL}^{\;-\dagger}\partial^{+}\psi_{TL}^{\;-} - \partial^{+}\psi_{TL}^{\;-\dagger}\psi_{TL}^{\;-} -$$

$$- \psi_{TR}{}^{+\dagger}\partial^{-}\psi_{TR}{}^{+} + \partial^{-}\psi_{TR}{}^{+\dagger}\psi_{TR}{}^{+} - \psi_{TR}{}^{-\dagger}\partial^{+}\psi_{TR}{}^{-} + \partial^{+}\psi_{TR}{}^{-\dagger}\psi_{TR}{}^{-} + \text{mass terms}] \tag{3.70}$$

After substituting for the various fields we find the *independent fields* (which create the in and out particle states) have the hamiltonian terms:

$$H = \sum_{\pm s} \int d^2p dp^+ \, p^- [b_{TL}{}^{+\dagger}(p,s) b_{TL}{}^{+}(p,s) - d_{TL}{}^{+}(p,s) d_{TL}{}^{+\dagger}(p,s) - - b_{TR}{}^{+\dagger}(p,s) b_{TR}{}^{+}(p,s) + d_{TR}{}^{+}(p,s) d_{TR}{}^{+\dagger}(p,s)] \tag{3.71}$$

$$= \sum_{\pm s} \int d^2p dp^+ \, p^- [b_{TL}{}^{+\dagger}(p,s) b_{TL}{}^{+}(p,s) + d_{TL}{}^{+\dagger}(p,s) d_{TL}{}^{+}(p,s) - - b_{TR}{}^{+\dagger}(p,s) b_{TR}{}^{+}(p,s) - d_{TR}{}^{+\dagger}(p,s) d_{TR}{}^{+}(p,s)] \tag{3.72}$$

up to the usual infinite constants due to left-handed operator rearrangement and right-handed operator rearrangement that are discarded. Eq. 3.72 is the basis for our particle interpretation of tachyon creation and annihilation operators based on Dirac's hole theory. Dirac hole theory as applied in light-front coordinates assumes all negative p^- ("energy") states are filled.

Left-Handed Tachyon Creation and Annihilation Operators

1. We identify $b_{TL}{}^{+\dagger}(p,s)$ and $d_{TL}{}^{+}(p,s)$ as creation operators for left-handed tachyons. $b_{TL}{}^{+\dagger}(p,s)$ creates a positive p^- ("energy") state and $d_{TL}{}^{+}(p,s)$ creates a negative p^- ("energy") state.

2. $b_{TL}{}^{+}(p,s)$ and $d_{TL}{}^{+\dagger}(p,s)$ are the corresponding annihilation operators for left-handed tachyons. $b_{TL}{}^{+}(p,s)$ annihilates a positive p^- ("energy") state and $d_{TL}{}^{+\dagger}(p,s)$ annihilates a negative p^- ("energy") state.

3. We assume Dirac hole theory holds for the left-handed tachyon vacuum with all negative energy states filled. There is no tachyon energy gap as there is for Dirac fermions. There is also the problem that the left-handed tachyon vacuum is not invariant under ordinary Lorentz transformations or Superluminal transformations. *However if we confine ourselves to light-front coordinates for computations no ambiguity can result and the Lorentz covariant quantities that we calculate, such as the S matrix, are well-defined.*

4. Using tachyon hole theory we identify $b_{TL}^{+}(p,s)$ and $d_{TL}^{+\dagger}(p,s)$ as annihilation operators for left-handed tachyons. $b_{TL}^{+}(p,s)$ annihilates a positive p^{-} ("energy") state and $d_{TL}^{+\dagger}(p,s)$ annihilates a negative p^{-} ("energy") state – thus creating a hole in the tachyon sea that we view as the creation of a positive p^{-} ("energy"), left-handed antitachyon. $d_{TL}^{+}(p,s)$ annihilates a positive p^{-} ("energy"), left-handed antitachyon.

Right-Handed Tachyons

The anti-commutation relations of right-handed tachyon creation and annhilation operators (eqs. 3.63) and the right-handed hamiltonian terms have the "wrong" sign compared to corresponding Dirac operators and left-handed tachyon operators. This situation is completely analogous to the situation of time-like photons in the covariant formulation of quantum Electrodynamics.[52] In the case of time-like photons it was possible to introduce an indefinite metric (Gupta-Bleuler formulation), and then to use the subsidiary condition $\partial A^{\nu}/\partial x^{\nu} = 0$ to reduce the dynamics of QED to the transverse components. Thus the time-like photons were intermediate artifacts needed to have a manifestly covariant formulation while QED observables depended solely on the transverse components of the electromagnetic field.

In the present case of free tachyons, and in leptonic ElectroWeak Theory there is no evident "subsidiary condition" to eliminate the right-handed tachyon fields. But

[52] Bogoliubov (1959) pp. 130-136.

since the only manner in which the right-handed leptonic tachyon fields[53] interact is through mass terms, which can be easily 'integrated out", right-handed leptonic tachyon fields are removed from the observable part of the leptonic ElectroWeak Theory by their "lack of interaction" with left-handed fields.

In the case of quark ElectroWeak Theory right-handed tachyon quark fields have charge (–1/3) and thus experience an electromagnetic interaction as well as a Z interaction. However, since quarks are totally confined, right-handed tachyon quarks will not be able to continuously emit photons or Z's due to energy conservation and their confinement to bound states of fixed positive energy. Starting in section 2.6, when we consider complex Lorentz group boosts, we will suggest that quarks may not consist of Dirac particles or tachyons of the type considered up to this point in this chapter. Rather they may be variants on Dirac particles and tachyons satisfying different dynamical equations. However, the preceding comments on quarks would still apply.

Thus right-handed tachyons are analogous to time-like photons – necessary theoretically but prevented from causing a negative energy disaster by the forms of their interactions. We discuss this subject in more detail in the following chapters.

3.5 Tachyon Feynman Propagator

In this section we develop the light-front propagator for tachyons. We begin with a subsection describing the light-front propagators of Dirac fields.

Dirac Field Light-Front Propagators

The light-front Feynman propagator for the ψ^+ field of a Dirac fermion is

$$iS^+{}_F(x,y)\gamma^0 = \theta(x^+ - y^+)\langle 0|\psi^+(x)\psi^{+\dagger}(y)|0\rangle - \theta(y^+ - x^+)\langle 0|\psi^{+\dagger}(y)\psi^+(x)|0\rangle \tag{3.73}$$

and does not contain a non-covariant piece due to the projection operators:

$$iS^+{}_F(x,y) = \int d^2p\,dp^+\theta(p^+)[1/(2(2\pi)^3p^+)]\{\theta(x^+ - y^+)[R^+(\not p + m)R^-]\,e^{-ip.(x-y)} +$$

[53] The tachyon fields are provisionally assumed to be neutrino fields in the leptonic sector, and d, s and b quarks in the quark sector.

$$+ \theta(y^+ - x^+)[R^+(-\not{p} + m)R^-]e^{+ip\cdot(x-y)}\}$$

$$= R^+ iS_F(x,y)R^- \qquad (3.74)$$

where $S_F(x,y)$ is the usual Feynman propagator.

The light-front Feynman propagator for a *left-handed* Dirac field ψ^+ is

$$iS^+_{LF}(x,y) = \int d^2p dp^+ \theta(p^+)[1/(2(2\pi)^3 p^+)]\{\theta(x^+ - y^+)[C^- R^+(\not{p} + m)R^- C^-]e^{-ip\cdot(x-y)} + $$
$$+ \theta(y^+ - x^+)[C^- R^+(-\not{p} + m)R^- C^-]e^{+ip\cdot(x-y)}\}$$

$$= C^- R^+ iS_F(x,y)R^- C^- \qquad (3.75)$$

Tachyon Field Feynman Propagator

Turning now to tachyons, the light-front Feynman propagator for the left-handed ψ_{TL}^+ *tachyon* field is (using the Fourier expansion of the left-handed tachyon field eqs. 3.56 and 3.57):

$$iS^+_{TLF}(x,y) = \theta(x^+ - y^+)\langle 0|\psi_{TL}^+(x)\psi_{TL}^{+\dagger}(y)\gamma^0|0\rangle -$$
$$- \theta(y^+ - x^+)\langle 0|\psi_{TL}^{+\dagger}(y)\gamma^0\psi_{TL}^+(x)|0\rangle$$
$$= -i\int d^2p dp^+ \theta(p^+) N_{TL}^{+2}(2m|\mathbf{p}|)^{-1} C^- R^+ \{\theta(x^+ - y^+)[(i\not{p} - m)\boldsymbol{\gamma}\cdot\mathbf{p}]e^{-ip\cdot(x-y)} +$$
$$+ \theta(y^+ - x^+)[(i\not{p} + m)\boldsymbol{\gamma}\cdot\mathbf{p}]e^{+ip\cdot(x-y)}\}R^+ C^- \gamma^0$$

If we define the on-shell momentum variable $p_0^- = (p_0^1 p_0^1 + p_0^2 p_0^2 - m^2)/(2p_0^+)$, $p_0^+ = p^+$, $p_0^j = p^j$ (for j = 1, 2), $p_{\perp 0}^2 = p_0^j p_0^j$ and $\not{p}_0 = p_0\cdot\gamma$ then the above equation can be rewritten as

$$= -iC^- R^+ \int d^4p[32\pi^4(p_0^+(p_0^+ - p_0^-) + p_{0\perp}^2)]^{-1} e^{-ip\cdot(x-y)} .$$

$$\cdot\{\theta(p^+)(i\not{p}_0 - m)\gamma\cdot\mathbf{p}_0]/[p^- - p_0^- + i\varepsilon] +$$

$$+ \theta(-p^+)(i\not{p}_0 + m)\gamma\cdot\mathbf{p}_0]/[p^- + p_0^- - i\varepsilon]\}R^+C^-\gamma^0$$

$$= -\tfrac{1}{2}\, i\int d^4p(2\pi)^{-4}[C^-R^+(i\not{p} - m)\gamma\cdot\mathbf{p}R^+C^-\gamma^0]e^{-ip.(x-y)}\cdot$$
$$\cdot[(p^2 + m^2 + i\varepsilon)(p^+(p^+ - p^-) + p_\perp^2))]^{-1}$$

and using $C^-R^+(i\not{p} - m)\gamma\cdot\mathbf{p}R^+C^- = i\, C^-R^+(p^+(p^+ - p^-) + p_\perp^2)$ we find

$$= \tfrac{1}{2}C^-R^+\gamma^0\int d^4p(2\pi)^{-4}\, p^+e^{-ip.(x-y)}/(p^2 + m^2 + i\varepsilon) \qquad (3.76)$$

Similarly the light-front Feynman propagator for the right-handed ψ_{TR}^+ tachyon field is

$$iS^+_{TRF}(x,y) = \theta(x^+ - y^+)\langle 0|\psi_{TR}^+(x)\psi_{TR}^{+\dagger}(y)\gamma^0|0\rangle -$$
$$- \theta(y^+ - x^+)\langle 0|\psi_{TR}^{+\dagger}(y)\gamma^0\psi_{TR}^+(x)|0\rangle$$

$$= -\tfrac{1}{2}C^+R^+\gamma^0\int d^4p(2\pi)^{-4}\, p^+e^{-ip.(x-y)}/(p^2 + m^2 + i\varepsilon) \qquad (3.77)$$

where the relative minus sign between eqs. 3.76 and 3.77 is due to the relative minus signs of the Fouier component operator anti-commutation relations in eq. 3.58 and 3.63. Thus we find *tachyon* pole terms in the tachyon propagator as one would expect.

3.6 Complex Lorentz Group L_C Spin ½ Fermion Equations and Wave Functions

In this section we consider L_C group boosts and use them to develop a wider set of dynamical equations for free spin ½ fermions.[54] In chapter 2 we defined L_C boosts with

$$\Lambda_C(\mathbf{v_c}) = \exp[i\omega\hat{\mathbf{w}}\cdot\mathbf{K}] \quad (2.61)$$
$$\omega = (\omega_r^{\,2} - \omega_i^{\,2} + 2i\omega_r\omega_i\,\hat{\mathbf{u}}_r\cdot\hat{\mathbf{u}}_i)^{1/2} \quad (2.62)$$
$$\hat{\mathbf{w}} = (\omega_r\hat{\mathbf{u}}_r + i\omega_i\hat{\mathbf{u}}_i)/\omega \quad (2.63)$$
$$\hat{\mathbf{w}}\cdot\hat{\mathbf{w}} = \hat{\mathbf{u}}_r\cdot\hat{\mathbf{u}}_r = \hat{\mathbf{u}}_i\cdot\hat{\mathbf{u}}_i = 1 \quad (2.64a)$$
$$\mathbf{v_c} = \hat{\mathbf{w}}\tanh(\omega) \quad (2.64b)$$

3.7 L_C Spinor "Normal" Lorentz Boosts & More Spin ½ Particle Types

Spinor boost transformations of extended Lorentz groups were used in sections 3.2 and 3.3 to develop the dynamical equations for Dirac fields and tachyon fields. In this section we will use L_C group spinor boosts to generate additional particle field dynamical equations.

The form of the L_C spinor boost transformation corresponding to the coordinate transformation eq. 2.61 is:

$$S_C(\omega, \mathbf{v_c}) = \exp(-i\omega\sigma_{0k}\hat{w}_k/2) = \exp(-\omega\gamma^0\boldsymbol{\gamma}\cdot\hat{\mathbf{w}}/2)$$

$$= \cosh(\omega/2)I + \sinh(\omega/2)\gamma^0\boldsymbol{\gamma}\cdot\hat{\mathbf{w}} \quad (3.78)$$

The inverse transformation is

$$S_C^{-1}(\omega, \mathbf{v_c}) = \gamma^2\gamma^0 K^{-1} S_C^{\dagger} K\gamma^0\gamma^2 = \gamma^2\gamma^0 S_C^{\,T}\gamma^0\gamma^2 = \exp(\omega\gamma^0\boldsymbol{\gamma}\cdot\hat{\mathbf{w}}/2)$$

[54] The complexon theory that we develop and use for quark dynamics in the Standard Model is not required. Our Standard Model could use Dirac fermion dynamics for the up-type quarks and tachyon dynamics for down-type quarks. We choose to use complexon dynamics for quarks because they have an internal SU(3)-like structure suggestive of color SU(3). More importantly, their spin dynamics is different and thus may resolve the differences between theory and experiment for the deep inelastic parton spin-dependent structure functions.

$$= \cosh(\omega/2)I - \sinh(\omega/2)\gamma^0\boldsymbol{\gamma}\cdot\hat{\mathbf{w}} \tag{3.79}$$

where the superscript T denotes the transpose and K is the complex conjugation operator (that also appears in the time-reversal operator). Note that S_C is not unitary just as in previous cases considered in this chapter.

We now redo the development of spin ½ dynamical equations of motion of sections 3.2 and 3.3 in this more general case of complex ω and $\hat{\mathbf{w}}$. Again we apply a boost to a Dirac equation for a positive energy plane wave particle of mass m at rest:

$$0 = S_C(\omega, \mathbf{v}_c))(m\gamma^0 - m)e^{-imt}w(0)$$

$$= [mS_C\gamma^0 S_C^{-1} - m]e^{-imt}S_C w(0) \tag{3.80}$$

where $S_C = S_C(\omega, \hat{\mathbf{w}})$. After some algebra

$$mS_C\gamma^0 S_C^{-1} = m[\cosh(\omega)\gamma^0 - \sinh(\omega)\boldsymbol{\gamma}\cdot\hat{\mathbf{w}}] \tag{3.81}$$

Case 1: Parallel Real and Imaginary Relative Vector

If the real and imaginary relative vectors parts of $\hat{\mathbf{w}}$, namely $\hat{\mathbf{u}}_r$ and $\hat{\mathbf{u}}_i$, are parallel, then $\hat{\mathbf{u}}_r\cdot\hat{\mathbf{u}}_i = 1$ and

$$\omega = \omega_r + i\omega_i \tag{3.82}$$

Eq. 2.64b enables us to re-express eq. 3.81 as

$$mS_C\gamma^0 S_C^{-1} = m[\cosh(\omega_r)\cos(\omega_i) + i\sinh(\omega_r)\sin(\omega_i)]\gamma^0 -$$

$$- m[\sinh(\omega_r)\cos(\omega_i) + i\cosh(\omega_r)\sin(\omega_i)]\boldsymbol{\gamma}\cdot\hat{\mathbf{u}}_r \tag{3.83}$$

or

$$mS_C\gamma^0 S_C^{-1} = \cos(\omega_i)\gamma\cdot p_r + i\sin(\omega_i)\gamma\cdot p_i \qquad (3.84)$$

where

$$p_r^0 = m\cosh(\omega_r) \qquad p_i^0 = m\sinh(\omega_r) \qquad (3.85)$$

and

$$\mathbf{p_r} = m\hat{\mathbf{u}}_r \sinh(\omega_r) \qquad \mathbf{p_i} = m\hat{\mathbf{u}}_r \cosh(\omega_r) \qquad (3.86)$$

If $\omega_i = 0$, then we recover the momentum space Dirac equation eq. 3.8. If $\omega_i = \pi/2$, then we obtain the left-handed momentum space tachyon equation eq. 3.16. Since the range of ω_i is [0, ∞> (due to the cut along the real ω-plane axis) eq. 3.84 corresponds to the results of the Left-Handed Extended Lorentz group development discussed earlier.

Case 2: Anti-Parallel Real and Imaginary Relative Vector

If the real and imaginary relative vectors parts of $\hat{\mathbf{w}}$, $\hat{\mathbf{u}}_r$ and $\hat{\mathbf{u}}_i$, are anti-parallel $\hat{\mathbf{u}}_r = -\hat{\mathbf{u}}_i$, then $\hat{\mathbf{u}}_r\cdot\hat{\mathbf{u}}_i = -1$ and

$$\omega = \omega_r - i\omega_i \qquad (3.87)$$

We can then express eq. 3.81 as

$$mS_C\gamma^0 S_C^{-1} = m[\cosh(\omega_r)\cos(\omega_i) - i\sinh(\omega_r)\sin(\omega_i)]\gamma^0 -$$

$$- m[\sinh(\omega_r)\cos(\omega_i) - i\cosh(\omega_r)\sin(\omega_i)]\boldsymbol{\gamma}\cdot\hat{\mathbf{u}}_r \qquad (3.88)$$

or

$$mS_C\gamma^0 S_C^{-1} = \cos(\omega_i)\gamma\cdot p_r - i\sin(\omega_i)\gamma\cdot p_i \qquad (3.89)$$

where

$$p_r^0 = m\cosh(\omega_r) \qquad p_i^0 = m\sinh(\omega_r) \qquad (3.90)$$

and

$$\mathbf{p_r} = m\hat{\mathbf{u}}_r \sinh(\omega_r) \qquad\qquad \mathbf{p_i} = m\hat{\mathbf{u}}_r \cosh(\omega_r) \tag{3.91}$$

If $\omega_i = 0$, then we again recover the momentum space Dirac equation eq. 3.8. If $\omega_i = \pi/2$, then we obtain the right-handed momentum space tachyon equation. (The range of ω_i is again $[0, \infty>$.)

Note: Since the matrix elements in the boost eq. 2.65 depend on $\gamma = (1 - \beta^2)^{-½}$ with a singularities at $\beta = \pm 1$, which in turn corresponds to $\omega = \pm\infty$, there is a branch cut along the ω axis in the complex ω-plane. Therefore we point out again the product of three Left-handed extended Lorentz transformations is *not* equivalent to a Right-handed extended Lorentz transformation.

Case 3: Complexons: A New Type of Particle with Perpendicular Real and Imaginary 3-Momenta

If the real and imaginary relative vectors parts of $\hat{\mathbf{w}}$, namely $\hat{\mathbf{u}}_r$ and $\hat{\mathbf{u}}_i$, are perpendicular, $\hat{\mathbf{u}}_r \cdot \hat{\mathbf{u}}_i = 0$, then by eq. 2.62

$$\omega = (\omega_r^2 - \omega_i^2)^{½} \tag{3.92}$$

Thus ω is either pure real ($\omega_r \geq \omega_i$) or pure imaginary ($\omega_r < \omega_i$).

The momentum space equation generated by the corresponding L_C spinor boost is

$$\{m \cosh(\omega)\gamma^0 - m \sinh(\omega)\boldsymbol{\gamma}\cdot(\omega_r\hat{\mathbf{u}}_r + i\omega_i\hat{\mathbf{u}}_i)/\omega - m\}e^{-ip.x}w_c(p) = 0 \tag{3.93}$$

Defining the momentum 4-vector

$$p = (p^0, \mathbf{p}) \tag{3.94}$$

where

$$p^0 = m \cosh(\omega) \qquad\qquad \mathbf{p} = \mathbf{p_r} + i\mathbf{p_i} \tag{3.95}$$

with

$$\mathbf{p}_r = m\omega_r\hat{\mathbf{u}}_r \sinh(\omega)/\omega \qquad \mathbf{p}_i = m\omega_i\hat{\mathbf{u}}_i \sinh(\omega)/\omega \tag{3.96a}$$

$$\mathbf{p}_r\cdot\mathbf{p}_i = 0 \tag{3.96b}$$

then eq. 3.93 becomes a positive energy Dirac-like equation

$$[p\cdot\gamma - m]e^{-ip.x}w_c(p) = 0$$

or, explicitly, (3.97)

$$[p^0\gamma^0 - (\mathbf{p}_r + i\mathbf{p}_i)\cdot\gamma - m]e^{-ip.x}w_c(p) = 0$$

with a complex 3-momentum $\mathbf{p}$ and the 4-momentum mass shell condition:

$$p^2 = p^{0\,2} - \mathbf{p}_r\cdot\mathbf{p}_r + \mathbf{p}_i\cdot\mathbf{p}_i = m^2 \tag{3.98}$$

Note

$$|\mathbf{v}| = |\mathbf{p}|/p^0 = [(\mathbf{p}_r + i\mathbf{p}_i)\cdot(\mathbf{p}_r + i\mathbf{p}_i)]^{1/2}/p^0 = \tanh(\omega) \tag{3.99}$$

and so the Lorentz factor

$$\gamma = \cosh(\omega) \tag{3.100}$$

Eq. 3.97 is the momentum space equivalent of the wave equation

$$[i\gamma^0\partial/\partial t + i\gamma\cdot(\nabla_r + i\nabla_i) - m]\psi_C(t, \mathbf{x}_r, \mathbf{x}_i) = 0 \tag{3.101}$$

where $\mathbf{x} = \mathbf{x}_r - i\mathbf{x}_i$, and where the grad operators ∇_r and ∇_i are with respect to $\mathbf{x}_r$ and $\mathbf{x}_i$ respectively. Since $\hat{\mathbf{u}}_r\cdot\hat{\mathbf{u}}_i = 0$, which in turn implies eq. 3.96b, we see that there is a subsidiary condition on the wave function

$$\nabla_r\cdot\nabla_i\, \psi_C(t, \mathbf{x}_r, \mathbf{x}_i) = 0 \tag{3.102a}$$

We will call the particles satisfying eqs. 3.101 and 3.102a *complexons*. In addition eq. 3.96b implies the anti-commutation relation

$$\{\boldsymbol{\gamma}\cdot\mathbf{p}_r, \boldsymbol{\gamma}\cdot\mathbf{p}_i\} = 0 \qquad (3.102b)$$

which in turn implies

$$\boldsymbol{\gamma}\cdot\nabla_r\boldsymbol{\gamma}\cdot\nabla_i\psi_C(t, \mathbf{x}_r, \mathbf{x}_i) = \boldsymbol{\gamma}\cdot\nabla_i\boldsymbol{\gamma}\cdot\nabla_r\psi_C(t, \mathbf{x}_r, \mathbf{x}_i) = 0 \qquad (3.102c)$$

also holds. We note that eq. 3.101 is covariant under the real Lorentz group and eq. 3.102 can be easily put into covariant form since the difference of these 4-vectors squared (which is real Lorentz group invariant): $[\gamma^0\partial/\partial t + \boldsymbol{\gamma}\cdot(\nabla_r + i\nabla_i)]^2 - [\gamma^0\partial/\partial t + i\boldsymbol{\gamma}\cdot(\nabla_r - i\nabla_i)]^2 = 4\nabla_r\cdot\nabla_i$.

Before considering a lagrangian formulation and the Fourier operator representation of $\psi_C(t, \mathbf{x}_r, \mathbf{x}_i)$ we will define the spinors and associated real and imaginary spin operators.

The spinor generated from a spin up Dirac spinor at rest by the complex Lorentz boost eqs. 3.78 is

$$w_c(p) = S_C(p)w(0) = [\cosh(\omega/2)I + \sinh(\omega/2)\gamma^0\boldsymbol{\gamma}\cdot\hat{\mathbf{w}}]w(0) \qquad (3.103)$$

Following a procedure similar to Appendix 3-A (which the reader may wish to examine first) we define four spinors for Dirac particles at rest:

$$w^k(0) = \begin{bmatrix} \delta_{1k} \\ \delta_{2k} \\ \delta_{3k} \\ \delta_{4k} \end{bmatrix} \qquad (3\text{-A}.2)$$

where Kronecker deltas appear in the brackets. Then by applying eq. 3.103 to the spinors defined by eq. 3-A.2 we find the L_C spinors

$$S_C w^k(0) = w_{Cr}{}^k(p) + i w_{Ci}{}^k(p) \qquad (3.104)$$

where

$$S_{Cr} = \cosh(\omega/2)I + (\omega_r/\omega)\sinh(\omega/2)\gamma^0\boldsymbol{\gamma}\cdot\hat{\mathbf{u}}_r$$

$$= [(m+E)/(2m)]^{1/2}I + [m(m+E)]^{-1/2}\gamma^0\boldsymbol{\gamma}\cdot\mathbf{p}_r = aI + b\gamma^0\boldsymbol{\gamma}\cdot\mathbf{p}_r \qquad (3.105)$$

by eqs. 3.95 and 3.96. Thus the "real" spinors $w_{Cr}{}^k(p)$ are the columns of

$$S_{Cr} = \begin{bmatrix} \underline{w_{Cr}{}^1(p)} & \underline{w_{Cr}{}^2(p)} & \underline{w_{Cr}{}^3(p)} & \underline{w_{Cr}{}^4(p)} \\ a & 0 & bp_{r\,z} & bp_{r-} \\ 0 & a & bp_{r+} & -bp_{r\,z} \\ bp_{r\,z} & bp_{r-} & a & 0 \\ bp_{r+} & -bp_{r\,z} & 0 & a \end{bmatrix} \qquad (3.106)$$

where $p_{r\pm} = p_{r\,x} \pm ip_{r\,y}$. The "imaginary" spinors are the columns of

$$S_{Ci} = (\omega_i/\omega)\sinh(\omega/2)\gamma^0\boldsymbol{\gamma}\cdot\hat{\mathbf{u}}_i = [m(m+E)]^{-1/2}\gamma^0\boldsymbol{\gamma}\cdot\mathbf{p}_i = b\gamma^0\boldsymbol{\gamma}\cdot\mathbf{p}_i \qquad (3.107)$$

$$S_{Ci} = \begin{bmatrix} \underline{w_{Ci}^{1}(p)} & \underline{w_{Ci}^{2}(p)} & \underline{w_{Ci}^{3}(p)} & \underline{w_{Ci}^{4}(p)} \\ 0 & 0 & bp_{i\,z} & bp_{i-} \\ 0 & 0 & bp_{i+} & -bp_{i\,z} \\ bp_{i\,z} & bp_{i-} & 0 & 0 \\ bp_{i+} & -bp_{i\,z} & 0 & 0 \end{bmatrix} \tag{3.108}$$

where $p_{i\pm} = p_{i\,x} \pm ip_{i\,y}$.

Eqs. 3.101 through 3.108 imply that the wave function solution of eq. 3.101, subject to the subsidiary condition eq. 102a, is[55, 56]

$$\psi_C(x_r, x_i) = \sum_{\pm s} \int d^3p_r d^3p_i\, N_C(p)\delta(\mathbf{p}_r\cdot\mathbf{p}_i/m^2)[b_C(p,s)u_C(p, s)e^{-i(p.x + p^*.x^*)/2} + $$
$$+ d_C^{\dagger}(p,s)v_C(p, s)e^{+i(p.x + p^*.x^*)/2}] \tag{3.109}$$

where $\mathbf{p} = \mathbf{p}_r + i\mathbf{p}_i$ (eq. 3.95), $\mathbf{x} = \mathbf{x}_r - i\mathbf{x}_i$, $p\cdot x = p^0x^0 - \mathbf{p}\cdot\mathbf{x}$, and where we use

$$(p\cdot x + p^*\cdot x^*)/2 = p^0x^0 - \mathbf{p}_r\cdot\mathbf{x}_r - \mathbf{p}_i\cdot\mathbf{x}_i \tag{3.109a}$$

---

[55] Note that when $|\mathbf{p}_i| \geq |\mathbf{p}_r|$ (for imaginary $\omega = (\omega_r^2 - \omega_i^2)^{½}$) the 3-momentum becomes imaginary $\mathbf{p}\cdot\mathbf{p} < 0$. However, since we will be identifying confined quarks with this type of particle – much modified by a confining color quark interaction – the issue of an imaginary 3-momentum in the hypothetical free quark case becomes moot. We note the energy gap between positive and negative energy states disappears so $E = 0$ is possible. Thus real Lorentz transformations can mix positive and negative energy states. The solution is to do all calculations in the light-front frame as we do for tachyons. Then the mixing issue is resolved. In the present case we second quantize on the "time-front" for illustrative purposes.

[56] We scale $\mathbf{p}_r\cdot\mathbf{p}_i$ with m^2 in the delta function for convenience. In the case of a zero mass particle some other scale could be used.

in the exponentials in order to avoid divergences that would appear in the calculation of the equal-time commutator, the Feynman propagator and other quantities of interest after second quantization. Note that

$$(\nabla_r + i\nabla_i)e^{-i(p.x + p^*.x^*)/2} = i(\mathbf{p}_r + i\mathbf{p}_i)e^{-i(p.x + p^*.x^*)/2} \tag{3.109b}$$

and

$$(\nabla_r + i\nabla_i)e^{-ip^*.x^*} = 0 \tag{3.109c}$$

for all p.

The wave function's conjugate (the hermitean conjugate modified by letting $\mathbf{x}_i \rightarrow -\mathbf{x}_i$ in addition to hermitean conjugation) is

$$\psi_C^\dagger(x) = \psi_C^\dagger(x_r, -x_i) = \sum_{\pm s}\int d^3p_r d^3p_i\, \delta(\mathbf{p}_r\cdot\mathbf{p}_i/m^2)N_C(p^*)\cdot$$
$$\cdot[b_C^\dagger(p^*,s)u_C^\dagger(p^*,s)e^{+i(p.x^* + p^*.x)/2} + d_C(p^*,s)v_C^\dagger(p^*,s)e^{-i(p.x^* + p^*.x)/2}] \tag{3.110}$$

where $\mathbf{p} = \mathbf{p}_r + i\mathbf{p}_i$ (eq. 3.95), $\mathbf{x} = \mathbf{x}_r - i\mathbf{x}_i$, $p\cdot x = p^0x^0 - \mathbf{p}\cdot\mathbf{x}$, and † indicates hermitean hermitean conjugation.

The spinors are

$$\begin{aligned}
u_C(p, s) &= S_C(p)w^1(0)\\
u_C(p, -s) &= S_C(p)w^2(0)\\
v_C(p, s) &= S_C(p)w^3(0)\\
v_C(p, -s) &= S_C(p)w^4(0) \qquad\qquad (3.110a)\\
u_C^\dagger(p^*, s) &= w^{1T}(0)S_C^\dagger(p^*)= w^{1T}(0)S_C(p)\\
u_C^\dagger(p^*, -s) &= w^{2T}(0)S_C^\dagger(p^*) = w^{2T}(0)S_C(p)\\
v_C^\dagger(p^*, s) &= w^{3T}(0)S_C^\dagger(p^*) = w^{3T}(0)S_C(p)\\
v_C^\dagger(p^*, -s) &= w^{4T}(0)S_C^\dagger(p^*) = w^{4T}(0)S_C(p)
\end{aligned}$$

with the superscript "T" indicating the transpose. Note that

$$S_C^{\dagger}(p^*) = [S_C(p^*)]^{\dagger} = S_C(p) \tag{3.110b}$$

The normalization factor $N_C(p)$ is

$$N_C(p) = [2m/((2\pi)^6 p^0)]^{½} \tag{3.110c}$$

Since $\mathbf{p_r} = \mathbf{p_i} = 0$ in the particle rest frame prior to the complex group boost, the boosted particle spin 4-vector s^μ satisfies

$$s^\mu p_r^{\ \mu} = s^\mu p_i^{\ \mu} = 0 \tag{3.111}$$

Note that s^μ is itself complex[57] and, if the spin points in the z-direction prior to the complex boost (eq. 2.57), then the boosted s^μ has the form

$$s^\mu = (-\sinh(\omega)\hat{w}_z\,, (0,0,1) + (\cosh(\omega) - 1)\hat{w}_z\hat{\mathbf{w}}) \tag{3.112}$$

with $\hat{\mathbf{w}}$ defined by eq. 2.63: $\hat{\mathbf{w}} = (\omega_r\hat{\mathbf{u}}_r + i\omega_i\hat{\mathbf{u}}_i)/\omega = \mathbf{p}/(m\sinh(\omega))$ using eq. 3.96a.

A Global SU(3) Symmetry Revealed

Before proceding to consider the second quantization of this case, we will consider a global SU(3) symmetry implicit in eqs. 3.101, 3.102 and the solution 3.109. The defining property of the group SU(3) is that it preserves the invariance of inner products of complex 3-vectors of the form:

$$u^*\cdot v = u^{1*}v^1 + u^{2*}v^2 + u^{3*}v^3 \tag{3.113}$$

[57] This feature of partons, which is not present in ordinary Dirac particles, might be the source of the discrepancies between theory and experiment in deep inelastic parton spin physics which is based on conventional real parton spins.

If we examine the dynamical equation eq. 3.101 we see that the differential operator is invariant under an SU(3) transformation U (using $\nabla_c = (\nabla_c^*)^* = \mathbf{D}_c^*$)

$$[i\gamma^0\partial/\partial t + i\mathbf{D}_c^*\cdot\gamma - m] = [i\gamma^0\partial/\partial t + i\mathbf{D}_c'^*\cdot\gamma' - m] \qquad (3.114)$$

where

$$\mathbf{D}_c^* = \nabla_c = \nabla_r + i\nabla_i$$

and

$$\gamma'^a = U^{ab}\gamma'^b$$
$$D_c'^{*a} = D_c'^{*b}U^{\dagger ab}$$

where U is a global SU(3) transformation and $U^\dagger = U^{-1}$. By theorem[58] all 4 × 4 γ matrices such as **γ'** are equivalent up to a unitary transformation V. Thus $V^\dagger\gamma'V = \gamma$ and eq. 3.114 is equivalent to

$$[i\gamma^0\partial/\partial t + i\mathbf{D}_c^*\cdot\gamma - m] = [i\gamma^0\partial/\partial t + i\mathbf{D}_c'^*\cdot\gamma - m] \qquad (3.115)$$

$$= [i\gamma^0\partial/\partial t + i\nabla_c'\cdot\gamma - m]$$

where $\nabla_c'{}_a = U^{ab}\nabla_{cb}$ showing that eq. 3.101 is invariant under an SU(3) transformation if

$$\psi_C(t, \mathbf{x}) = \psi_C(t, U\mathbf{x}) = \psi_C'(t, \mathbf{x}') \qquad (3.116)$$

where $\psi_C(t, \mathbf{x}) \equiv \psi_C(t, \mathbf{x}_r, \mathbf{x}_i)$.

The subsidiary condition eq. 3.102a can be seen to transform as

$$\nabla_r\cdot\nabla_i\,\psi_C(t, \mathbf{x}) = \nabla_r^*\cdot\nabla_i\,\psi_C(t, \mathbf{x}) = \nabla_r'^*\cdot\nabla_i'\psi_C'(t, \mathbf{x}') = 0 \qquad (3.117)$$

under an SU(3) rotation. The invariance of the orthogonality condition is preserved.

---

[58] R. H. Good, Rev. Mod. Phys., **27**, 187 (1955).

The wave function (eq. 3.109) transforms in the following way under the SU(3) transformation U. If we define

$$q^{*\mu} = (q^0, \mathbf{q}^*) = (p^0, \mathbf{p}_r + i\mathbf{p}_i) = (p^0, \mathbf{p}) = p^\mu \qquad (3.118)$$

then eq. 3.109 can be rewritten in an almost[59] manifestly invariant form under a SU(3) transformation:

$$\psi_C(x) = \sum_{\pm s} \int d^3q_r d^3q_i\, N_C(p^0)\delta(\mathbf{q}_r^* \cdot \mathbf{q}_i/m^2)[b_C(q^*,s)u_C(q^*,s)e^{-i(q^*.x + q.x^*)/2} + $$
$$+ d_C^\dagger(q^*,s)v_C(q^*,s)e^{+i(q^*.x + q.x^*)/2}] \qquad (3.119)$$

subject to an examination of the transformation properties of the fourier coefficients and spinors. Note both terms in each exponential are separately invariant under global SU(3). (Note also $\mathbf{q}_r^* = \mathbf{q}_r$ since $\mathbf{q}_r$ is real.)

From the form of S_C (eq. 3.78) it is clear that an argument similar to that for the dynamical equations (eqs. 3.114 – 3.115) shows S_C is invariant under an SU(3) transformation and thus the spinors defined by eqs. 3.110a are also invariant under SU(3) transformations. The fourier coefficients, if second quantized in a direct generalization of the usual way, have covariant anti-commutation relations under an SU(3) transformation. For example

$$\{b_C(q,s), b_C^\dagger(q'^*,s')\} = \delta_{ss'}\delta^3(q_r - q'_{r'})\delta^3(q_i - q'_{i'}) \qquad (3.120)$$

Under an SU(3) transformation, z = Uq and z' = Uq', the right side of eq. 3.120 transforms to

---

[59] The δ-function in eq. 3.119, $\delta(\mathbf{q}_r^* \cdot \mathbf{q}_i/m^2)$, is not invariant under a global SU(3) transformation rather it is a constraint that breaks the SU(3) symmetry to O(3)×O(2) rather like a δ-function term in the path integral formalism of gauge theories "breaks" the symmetry and fixes the gauge. The O(3)×O(2) symmetry can be visualized by mapping (conceptually) the $\mathbf{q}_r$ and $\mathbf{q}_i$ to 3-space and considering the perpendicular vector **q** to the pair as rotating under O(3) transformations, and $\mathbf{q}_r$ and $\mathbf{q}_i$ rotating in the plane perpendicular to **q** under O(2) transformations.

$$\delta^3(q_r - q'_{r'})\delta^3(q_i - q'_{i'}) \rightarrow \delta^3(z_r - z'_{r'})\delta^3(z_i - z'_{i'})/|\partial(q)/\partial(z)|$$
$$= \delta^3(z_r - z'_{r'})\delta^3(z_i - z'_{i'}) \qquad (3.121)$$

where

$$|\partial(q)/\partial(z)| = |\partial(q_r^1,q_r^2,q_r^3,q_i^1,\, q_i^2,\, q_i^3)/\partial(z_r^1,z_r^2,z_r^3,z_i^1,\, z_i^2,\, z_i^3)| = 1 \qquad (3.122)$$

is the Jacobian of the transformation U. Thus the fourier coefficients transform trivially under SU(3). For example,

$$b_C(q^*,s) \rightarrow b_C(z^*,s) \qquad (3.123)$$

Since the integrand transforms as

$$\int d^3q_r d^3q_i \;\rightarrow \int d^3z_r d^3z_i\, |\partial(q)/\partial(z)| = \int d^3z_r d^3z_i \qquad (3.124)$$

the wave function $\psi_C(t, \mathbf{x})$ transforms as an SU(3) scalar (eq. 3.116) up to an inessential unitary transformation V of γ matrices: $\psi_C(t, \mathbf{x}) \rightarrow V\psi_C(t, \mathbf{x})$.[60]

Global SU(3) Spin ½ Complexon Fields

Having uncovered an SU(3) symmetry in the scalar field equations of Case 3A the generalization of the scalar field equations to the 3 representation of SU(3) is direct:

$$\psi_C{}^a(x) = \sum_{\pm s} \int d^3p_r d^3p_i\, N_C(p)\delta(\mathbf{p}_r\cdot\mathbf{p}_i/m^2)[b_C(p,a,s)u_C{}^a(p,\, s)e^{-i(p\cdot x\,+\,p^*\cdot x^*)/2} +$$
$$+\, d_C{}^\dagger(p,a,s)v_C{}^a(p,\, s)e^{+i(p\cdot x\,+\,p^*\cdot x^*)/2}] \qquad (3.125a)$$

---

[60] The spinors $u_C(q^*,s)$ and $v_C(q^*,s)$ are unchanged up to a unitary transformation of the γ matrices ($V^\dagger\gamma' V = \gamma$). Thus the term $(U\mathbf{w})^*\cdot\gamma' = \mathbf{w}^*\cdot V\gamma V^\dagger \equiv \mathbf{w}^*\cdot\gamma$ in the expressions for the $u_C(q^*,s)$ and $v_C(q^*,s)$ spinors.

for a = 1,2, 3 with $u_C{}^a(p, s)$ and $v_C{}^a(p, s)$ being the product a spinor of type eq. 3.110a and a 3 element column vector c^a with b^{th} element

$$c^a(b) = \delta^{ab} \tag{3.125b}$$

Under a global SU(3) transformation U the 3 complexon wave functions transform as

$$\psi_C{}'^a(x) = U^{ab}\psi_C{}^b(x) \tag{3.126}$$

In a subsequent chapter we will extend the global SU(3) symmetry described in these subsections to color local SU(3) upon the introduction of the Yang-Mills color gluon interaction and detach it from the global SU(3) coordinate transformation feature found in this section.

Lagrangian Formulation and Second Quantization of Complexons

In this subsection we will outline the canonical quantization of SU(3) singlet complexons with the quantum field equation

$$[i\gamma^0\partial/\partial t + i\gamma\cdot(\nabla_r + i\nabla_i) - m]\psi_C(t, \mathbf{x}_r, \mathbf{x}_i) = 0 \tag{3.101}$$

and subsidiary condition

$$\nabla_r\cdot\nabla_i\, \psi_C(t, \mathbf{x}_r, \mathbf{x}_i) = 0 \tag{3.102a}$$

We begin with the Lagrangian density

$$\mathcal{L} = \bar{\psi}_C(i\gamma^\mu D_\mu - m)\psi_C(x) \tag{3.127}$$

where $\bar{\psi}_C = \psi_C{}^\dagger\gamma^0$:

$$\psi_C{}^\dagger = [\psi_C(\mathbf{x}_r, \mathbf{x}_i)]^\dagger \Big|_{\mathbf{x}_i = -\mathbf{x}_i} \tag{3.128}$$

$$D_0 = \partial/\partial x^0$$
$$D_k = \partial/\partial x^k + i\,\partial/\partial x_i^{\,k} \qquad (3.129)$$

with $x^k = x_r^{\,k}$ for k = 1,2, 3. The invariant action (under real Lorentz transformations) is

$$I = \int d^7x\,\mathfrak{L} \qquad (3.130)$$

It is easy to show that the action is real

$$I^* = I \qquad (3.131)$$

in a manner similar to the case considered in Appendix 3-A due to the form of $\psi_C^{\,\dagger}$ in eq. 3.128. (One has to change the integration over $\mathbf{x_i}$ to $-\mathbf{x_i}$ after taking the complex conjugate of I and performing manipulations similar to those in Appendix 3-A.)

The conjugate momentum is

$$\pi_{Ca} = \partial\mathfrak{L}/\partial\dot{\psi}_{Ca} \equiv \partial\mathfrak{L}/\partial(\partial\psi_{Ca}/\partial x^0) = i\psi_{C\,a}^{\,\dagger} \qquad (3.132)$$

where a is a spinor index. It yields the non-zero anti-commutation relation

$$\{\psi_{C\,a}^{\,\dagger}(x), \psi_{Cb}(y)\} = \delta_{ab}\,\delta^3(x_r - y_r)\delta^3(x_i - y_i) \qquad (3.133a)$$

However we will see that the constraint eq. 3.102a is required. So the correct anti-commutator turns out to be

$$\{\psi_{C\,a}^{\,\dagger}(x), \psi_{Cb}(y)\} = -\delta_{ab}\delta'(\nabla_r\cdot\nabla_i/m^2)[\delta^3(x_r - y_r)\delta^3(x_i - y_i)] \qquad (3.133b)$$

where all ∇_r and ∇_i are ∇ derivatives with respect to x, and where $\delta'(\nabla_r\cdot\nabla_i)$ is the derivative of a delta function with the argument being differential operators such as those in eq. 3.102a. The minus sign is due to the presence of a *derivative* of a delta-function and not an issue.

The hamiltonian density is

$$\mathcal{H} = \pi_C \dot{\psi}_C - \mathcal{L} = \psi_C^{\dagger}(-i\boldsymbol{\alpha}\cdot\mathbf{D} + \beta m)\psi_C \quad (3.134)$$

and the (unsymmetrized) energy-momentum tensor is

$$\mathcal{T}_{\mu\nu} = -g_{\mu\nu}\mathcal{L} + \partial\mathcal{L}/\partial(D^{\mu}\psi_C)D_{\nu}\psi_C \quad (3.135)$$

The conserved energy and momentum are

$$P^0 = H = \int d^3x_r d^3x_i\, \mathcal{T}^{00} = \int d^3x_r d^3x_i\, \mathcal{H} \quad (3.136)$$

and

$$P^I = \int D^3X_R D^3X_I\, \mathcal{T}^{0I}$$

(3.137)

We now proceed to establish the canonical anti-commutation relations eq. 3.133. First, the second quantization of the complexon field (eqs. 3.109 and 3.110) uses the fourier coefficient anti-commutation relations eq. 3.120 (suitably rewritten):

$$\begin{aligned}
&\{b_C(p,s), b_C^{\dagger}(p'^*,s')\} = \delta_{ss'}\delta^3(\mathbf{p_r} - \mathbf{p'_{r'}})\delta^3(\mathbf{p_i} + \mathbf{p'_{i'}})\\
&\{d_C(p,s), d_C^{\dagger}(p'^*,s')\} = \delta_{ss'}\delta^3(\mathbf{p_r} - \mathbf{p'_{r'}})\delta^3(\mathbf{p_i} + \mathbf{p'_{i'}})\\
&\{b_C(p,s), b_C(p'^*,s')\} = \{d_C(p,s), d_C(p'^*,s')\} = 0\\
&\{b_C^{\dagger}(p,s), b_C^{\dagger}(p'^*,s')\} = \{d_C^{\dagger}(p,s), d_C^{\dagger}(p'^*,s')\} = 0\\
&\{b_C(p,s), d_C^{\dagger}(p'^*,s')\} = \{d_C(p,s), b_C^{\dagger}(p'^*,s')\} = 0\\
&\{b_C^{\dagger}(p,s), d_C^{\dagger}(p'^*,s')\} = \{d_C(p,s), b_C(p'^*,s')\} = 0
\end{aligned} \quad (3.138)$$

The delta-function arguments $\delta^3(\mathbf{p}_i + \mathbf{p'}_{i'})$ above have a positive sign in order to obtain $\delta^3(\mathbf{x}_i - \mathbf{y}_i)$ in the field anti-commutator eq. 3.133b.

The spinors, eq. 3.110a, satisfy

$$\sum_{\pm s} u_\alpha(p, s)\bar{u}_\beta(p^*, s) = (2m)^{-1}(\not{p} + m)_{\alpha\beta} \quad (3.139)$$
$$\sum_{\pm s} v_\alpha(p, s)\bar{v}_\beta(p^*, s) = (2m)^{-1}(\not{p} - m)_{\alpha\beta}$$

by eqs. 3.78, 3.81, 3.85 and 3.103 remembering

$$\bar{u}_C(p^*,s) = w^{1T}(0)S_C(p)\gamma^0 = w^{1T}(0)[\cosh(\omega/2)I + \sinh(\omega/2)\gamma^0\boldsymbol{\gamma}\cdot\hat{\mathbf{w}}]\gamma^0 \quad (3.140)$$

by eqs. 3.110 since $\hat{\mathbf{w}}^{**} = \hat{\mathbf{w}}$.

We will now evaluate the equal-time anti-commutation relation using eqs. 3.109 and 3.110:

$$\{\psi_{C\,a}^{\dagger}(x), \psi_{Cb}(y)\} = \sum_{\pm s,\, s'} \int d^3p_r d^3p_i\, d^3p'_r d^3p'_i\, \delta(\mathbf{p}_r\cdot\mathbf{p}_i/m^2)\delta(\mathbf{p}'_r\cdot\mathbf{p}'_i/m^2)\, N_C(p')N_C(p)\cdot$$
$$[\{b_C^{\dagger}(p^*,s)u_{Ca}^{\dagger}(p^*,s)e^{+i(p.x^* + p^*.x)/2},\ b_C(p',s')u_{Cb}(p', s')e^{-i(p'.y + p'^*.y^*)/2}\}+$$
$$+ \{d_C(p^*,s)v_{Ca}^{\dagger}(p^*,s)e^{-i(p.x^* + p^*.x)/2},\ d_C^{\dagger}(p',s')v_{Cb}(p', s')e^{+i(p'.y + p'^*.y^*)/2}\}]$$

$$= \int d^3p_r d^3p_i\, N_C^2(p)[\delta(\mathbf{p}_r\cdot\mathbf{p}_i/m^2)]^2[((\not{p} + m)\gamma^0)_{ba}e^{+i(p.x^* + p^*.x)/2 - i(p^*.y + p.y^*)/2} +$$
$$+((\not{p} - m)\gamma^0)_{ba}e^{-i(p.x^* + p^*.x)/2 + i(p^*.y + p.y^*)/2}]/(2m)$$

Next we use eq. 3.110c and the identity

$$[\delta(x - y)]^2 = -\tfrac{1}{2}\,\delta'(x - y) \equiv -\tfrac{1}{2}\, d\delta(x - y)/dx \quad (3.141)$$

which can be derived from the step function identity $\theta(x - y) = [\theta(x - y)]^2$ to obtain

$$\{\psi_{C\,a}^{\dagger}(x),\psi_{Cb}(y)\} = -\tfrac{1}{2}\int d^3p_r d^3p_i N_C^2(p)\delta'(\mathbf{p}_r\cdot\mathbf{p}_i/m^2)[((\not{p}+m)\gamma^0)_{ba} e^{-i\mathbf{p}_r\cdot(\mathbf{x}_r-\mathbf{y}_r)+i\mathbf{p}_i\cdot(\mathbf{x}_i-\mathbf{y}_i)} +$$

$$+ ((\not{p}-m)\gamma^0)_{ba} e^{+i\mathbf{p}_r\cdot(\mathbf{x}_r-\mathbf{y}_r)-i\mathbf{p}_i\cdot(\mathbf{x}_i-\mathbf{y}_i)}]/(2m)$$

$$= -\tfrac{1}{2}\delta_{ba}\int d^3p_r d^3p_i N_C^2(p)\delta'(\mathbf{p}_r\cdot\mathbf{p}_i/m^2)p^0 e^{-i\mathbf{p}_r\cdot(\mathbf{x}_r-\mathbf{y}_r)+i\mathbf{p}_i\cdot(\mathbf{x}_i-\mathbf{y}_i)}/m$$

$$= -\delta_{ab}\,\delta'(\nabla_r\cdot\nabla_i/m^2)[\delta^3(x_r - y_r)\delta^3(x_i - y_i)] \qquad (3.142)$$

The grad operators, ∇_r and ∇_i, are derivatives are with respect to x of the Dirac delta functions. The factor[61] $\delta'(\nabla_r\cdot\nabla_i)$ expresses the orthogonality constraint in coordinate space on the momenta eq. 3.102a. It is analogous to the transversality constraint on the electromagnetic vector potential commutator:

$$[\pi_A^{\,j}(x), A_k(y)] = -i\,\delta^{tr}_{jk}(x - y) \qquad (3.143)$$

$$\delta^{tr}_{jk}(x - y) = (\delta_{jk} - \partial_j\partial_k/\nabla^2)\,\delta^3(x - y) \qquad (3.144)$$

where $\partial_k = \partial/\partial x_k$.

Complexon Feynman Propagator

The complexon Feynman propagator for ψ_C is[62]

$$iS_C(x, y) = \theta(x^0 - y^0)\langle 0|\psi_C(x)\psi_C^{\dagger}(y)\gamma^0|0\rangle -$$

---

[61] A derivative of a delta function containing grad operators.

[62] The reader, upon seeing the additional integrations d^3p_i might suspect that they would ultimately lead to divergence issues in perturbation theory calculations. However the $\delta'(\mathbf{p}_r\cdot\mathbf{p}_i/m^2)$ term compensates in part for the additional integrations by four powers of momentum since $\delta'(\mathbf{p}_r\cdot\mathbf{p}_i/m^2) = (|\mathbf{p}_r||\mathbf{p}_i|/m^2)^{-2}\delta'(\cos\theta_{ri})$ where θ_{ri} is the angle between the momenta. As a result only 2 fermion and 3 fermion loop integrations would potentially have difficulties if one uses the conventional approach to perturbation theory. If one uses the approach of Blaha (2003) and (2005a) then there are no divergences.

$$- \theta(y^0 - x^0)\langle 0|\psi_C{}^\dagger(y)\gamma^0\psi_C(x)|0\rangle \qquad (3.145)$$

$$= \int d^3p_r d^3p_i N_C{}^2(p)[\delta(\mathbf{p}_r\cdot\mathbf{p}_i/m^2)]^2\{\theta(x^0 - y^0)(\not{p} + m)e^{-i(p^*\cdot(x-y) + p\cdot(x^*-y^*))/2} -$$
$$- \theta(y^0 - x^0)(\not{p} - m)e^{+i(p^*\cdot(x-y) + p\cdot(x^*-y^*))/2}\}/(2m)$$

$$= -(4\pi)^{-1}\int dp^0 d^3p_r d^3p_i (2\pi)^{-6}\delta'(\mathbf{p}_r\cdot\mathbf{p}_i/m^2)(\not{p} + m)e^{-i(p^*\cdot(x-y) + p\cdot(x^*-y^*))/2}/(p^2 - m^2 + i\varepsilon)$$

$$= -\tfrac{1}{2}\int dp^0 d^3p_r d^3p_i\, \delta'(\mathbf{p}_r\cdot\mathbf{p}_i/m^2)(\not{p} + m)(2\pi)^{-7}\cdot$$
$$\cdot\exp[-ip^0(x^0 - y^0) + i\mathbf{p}_r\cdot(\mathbf{x}_r - \mathbf{y}_r) - i\mathbf{p}_i\cdot(\mathbf{x}_i - \mathbf{y}_i)]/(p^2 - m^2 + i\varepsilon) \qquad (3.146)$$

The integral can be written in the form:

$$I = \int dp^0 d^3p_r d^3p_i \delta'(\mathbf{p}_r\cdot\mathbf{p}_i/m^2)(\not{p} + m)\exp[-ip^0(x^0 - y^0) + i\mathbf{p}_r\cdot(\mathbf{x}_r - \mathbf{y}_r) - i\mathbf{p}_i\cdot(\mathbf{x}_i - \mathbf{y}_i)]/(p^2 - m^2 + i\varepsilon)$$

$$= \int d^4p_r dM^2\delta'(\nabla_r\cdot\nabla_i/m^2)(p^0\gamma^0 - (\mathbf{p}_r - \nabla_i)\cdot\gamma + m)\exp[-ip^0(x^0 - y^0) + i\mathbf{p}_r\cdot(\mathbf{x}_r - \mathbf{y}_r)]\cdot$$
$$\cdot J(\mathbf{x}_i - \mathbf{y}_i, M^2)/(p_r{}^2 - M^2 + i\varepsilon) \qquad (3.147)$$

where $p_r{}^2 = p^{0\,2} - \mathbf{p}_r\cdot\mathbf{p}_r$ and

$$J(\mathbf{x}_i - \mathbf{y}_i, M^2) = (2\pi)^{-3}\int d^3p_i\, \delta(M^2 + \mathbf{p}_i{}^2 - m^2)\exp[-i\mathbf{p}_i\cdot(\mathbf{x}_i - \mathbf{y}_i)] \qquad (3.148)$$

$$= (2\pi)^{-2}|\mathbf{x}_i - \mathbf{y}_i|^{-1}\theta(m^2 - M^2)\sin((m^2 - M^2)^{1/2}|\mathbf{x}_i - \mathbf{y}_i|)$$

The complexon Feynman propagator eq. 3.146 can be rearranged into the form of a spectral integral:

$$iS_C(x, y) = -\int dM\, (i\gamma^0\partial/\partial x^0 - i(\nabla_r - i\nabla_i)\cdot\gamma + m)\delta'(\nabla_r\cdot\nabla_i/m^2)\cdot$$

$$\cdot J(\mathbf{x_i} - \mathbf{y_i}, M^2)\Delta_F(x - y, M) \quad (3.149)$$

where

$$\Delta_F(x - y, M) = (2\pi)^{-4}\int d^4p_r \exp[-ip^0(x^0 - y^0) + i\mathbf{p_r}\cdot(\mathbf{x_r} - \mathbf{y_r})]/(p_r^2 - M^2 + i\varepsilon) \quad (3.150)$$

Case 4: Left-handed Tachyon Complexons

In this case $\hat{\mathbf{u}}_\mathbf{r}\cdot\hat{\mathbf{u}}_\mathbf{i} = 0$ again. However we add an imaginary term to ω to obtain a manifest Left-handed L_C boost[63]

$$\Lambda_{CL}(\mathbf{v_c}) = \exp[i(\omega + i\pi/2)\hat{\mathbf{w}}\cdot\mathbf{K}] \quad (3.151)$$

where ω remains

$$\omega = (\omega_r^2 - \omega_i^2)^{1/2} \quad (2.62)$$

and

$$\hat{\mathbf{w}} = (\omega_r\hat{\mathbf{u}}_\mathbf{r} + i\omega_i\hat{\mathbf{u}}_\mathbf{i})/\omega \quad (2.63)$$

$$\hat{\mathbf{w}}\cdot\hat{\mathbf{w}} = \hat{\mathbf{u}}_\mathbf{r}\cdot\hat{\mathbf{u}}_\mathbf{r} = \hat{\mathbf{u}}_\mathbf{i}\cdot\hat{\mathbf{u}}_\mathbf{i} = 1 \quad (2.64a)$$

$$\mathbf{v_c} = \hat{\mathbf{w}}\tanh(\omega + i\pi/2) = \hat{\mathbf{w}}\,\mathrm{cotanh}(\omega) \quad (3.152)$$

Letting $\omega_L = \omega + i\pi/2$ we find, as before,

$$\cosh(\omega_L) = i\sinh(\omega) = -\gamma = i\,\gamma_s$$
$$\sinh(\omega_L) = i\cosh(\omega) = -\beta\gamma = i\beta\gamma_s \quad (2.11)$$

with, $\beta = v_c = |\mathbf{v_c}| > 1$, $\gamma_s = (\beta^2 - 1)^{-1/2}$, and

[63] The reader can readily verify the form is consistent with an L_C boost transformation.

$$\sinh(\omega) = \gamma_s$$
$$\cosh(\omega) = \beta\gamma_s \tag{2.12}$$

Thus we denote $\Lambda_{CL}(\mathbf{v}_c)$ by

$$\Lambda_{CL}(\mathbf{v}_c) \equiv \Lambda_{CL}(\omega, \hat{\mathbf{w}}) \tag{3.153}$$

The corresponding spinor boost transformation is:

$$S_{CL}(\Lambda_{CL}(\omega, \hat{\mathbf{w}})) = \exp(-i\omega_L\sigma_{0i}\hat{w}_i/2) = \exp(-\omega_L\gamma^0\boldsymbol{\gamma}\cdot\hat{\mathbf{w}}/2)$$
$$= \cosh(\omega_L/2)I + \sinh(\omega_L/2)\gamma^0\boldsymbol{\gamma}\cdot\hat{\mathbf{w}} \tag{3.154}$$

The momentum space equation generated by $S_{CL}(\Lambda_{CL}(\omega, \hat{\mathbf{w}}))$ is

$$\{m\cosh(\omega_L)\gamma^0 - m\sinh(\omega_L)\boldsymbol{\gamma}\cdot(\boldsymbol{\omega_r}\hat{\mathbf{u}}_r + i\boldsymbol{\omega_i}\hat{\mathbf{u}}_i)/\omega - m\}e^{+ip.x}w_{cL}(p) = 0 \tag{3.155}$$

or

$$\{im\sinh(\omega)\gamma^0 - im\cosh(\omega)\boldsymbol{\gamma}\cdot(\boldsymbol{\omega_r}\hat{\mathbf{u}}_r + i\boldsymbol{\omega_i}\hat{\mathbf{u}}_i)/\omega - m\}e^{+ip.x}w_{cL}(p) = 0 \tag{3.156}$$

where $p\cdot x = Et - \mathbf{p}\cdot\mathbf{x}$ after performing a corresponding left-handed superluminal coordinate transformation in the exponential factor based on eq. 2.10d. Thus the positive energy wave is transformed into a negative energy wave by the transformation.

The momentum 4-vector is defined by

$$p = (p^0, \mathbf{p}) \tag{3.157}$$

where

$$p^0 = m\sinh(\omega) \qquad \mathbf{p} = \mathbf{p}_r + i\mathbf{p}_i \tag{3.158}$$

with

$$\mathbf{p_r} = m\omega_r\hat{\mathbf{u}}_r \cosh(\omega)/\omega \qquad \mathbf{p_i} = m\omega_i\hat{\mathbf{u}}_i \cosh(\omega)/\omega \tag{3.159}$$

and

$$\mathbf{p_r}\cdot\mathbf{p_i} = 0 \tag{3.160}$$

then eq. 3.156 becomes the complexon tachyon equation

$$[ip\cdot\gamma - m]e^{+ip.x}w_{cL}(p) = 0 \tag{3.161}$$

with a complex 3-momentum **p** and the tachyon 4-momentum mass shell condition:[64]

$$p^2 = p^{0\,2} - \mathbf{p_r}^2 + \mathbf{p_i}^2 = -m^2 \tag{3.162}$$

Eq. 3.161 is the momentum space equivalent of the wave equation

$$[\gamma^0\partial/\partial t + \boldsymbol{\gamma}\cdot(\nabla_r + i\nabla_i) - m]\psi_{CL}(t, \mathbf{x_r}, \mathbf{x_i}) = 0 \tag{3.163a}$$

or

$$[\gamma\cdot\nabla - m]\psi_{CL}(t, \mathbf{x_r}, \mathbf{x_i}) = 0 \tag{3.163b}$$

with the subsidiary condition on the wave function

$$\nabla_r\cdot\nabla_i\, \psi_{CL}(t, \mathbf{x_r}, \mathbf{x_i}) = 0 \tag{3.164}$$

also holds. We note that eq. 3.163 is covariant under the real Lorentz group and eq. 3.164 can be easily put into (real Lorentz group) covariant form.

Before considering a lagrangian formulation and the Fourier operator representation of $\psi_{CL}(t, \mathbf{x_r}, \mathbf{x_i})$ we will define the tachyon spinors, and its associated real and imaginary spin operators.

---

[64] Note that the presence of the $\mathbf{p_i}^2$ term does not change the tachyon requirement that $\mathbf{p_r}^2 \geq m^2$ as seen in the previous cases.

The spinor generated from a spin up Dirac spinor at rest by the L_C spinor boost eqs. 3.154 is

$$w_{cL}(p) = S_{CL}w(0) = [\cosh(\omega_L/2)I + \sinh(\omega_L/2)\gamma^0\boldsymbol{\gamma}\cdot\hat{\mathbf{w}}]w(0) \quad (3.165)$$

Following a procedure similar to Appendix 3-A (which the reader may wish to examine first) we define four spinors for Dirac particles at rest with eq. 3-A.2. Then by applying a boost to these rest spinors we find the L_C tachyon spinors:

$$S_{CL}w^k(0) = w_{cL}{}^k(p) \quad (3.166)$$

and from these tachyon spinors we generalize to tachyon spinors $u_{CL}(p, s)$ and $v_{CL}(p, s)$ in a manner similar to eqs. 3.110a of the previous case.

Eqs. 3.161 through 3.164 imply that the wave function solution of eq. 3.161, subject to the subsidiary condition eq. 3.164, has the form

$$\psi_{CL}(x) = \sum_{\pm s} \int_{\mathbf{p}_r^2 \geq m^2} d^3p_r d^3p_i\, N_{CL}(p)\delta(\mathbf{p}_r\cdot\mathbf{p}_i/m^2)[b_{CL}(p,s)u_{CL}(p, s)e^{-i(p\cdot x + p^*\cdot x^*)/2} +$$

$$d_{CL}{}^\dagger(p,s)v_{CL}(p, s)e^{+i(p\cdot x + p^*\cdot x^*)/2}] \quad (3.167)$$

where $\mathbf{p} = \mathbf{p}_r + i\mathbf{p}_i$, $\mathbf{x} = \mathbf{x}_r - i\mathbf{x}_i$, $p\cdot x = p^0x^0 - \mathbf{p}\cdot\mathbf{x}$, and $b_{CL}(p, s)$ and $d_{CL}(p,s)$ are tachyon fourier coefficients.

Global SU(3) Symmetry

We can show that there is also a global SU(3) symmetry present here as shown in the previous case. The demonstration is similar to that of eqs. 3.113 – 3.126.

Light-Front Quantization of Tachyonic Complexons

Because of the momentum constraint $\mathbf{p}_r^2 \geq m^2$ the set of solutions of the form of eq. 3.167 is incomplete and the result of second quantization will not be an equal time anti-commutator expression consisting of derivatives of delta functions (eq. 3.142) but rather an analogue to previous unsuccessful attempts to create a second quantized tachyon theory.[65]

Therefore we will use light-front coordinates, and left and right handed field operators (as previously) to obtain a successful second quantization of this new type of tachyon.

The "missing" factor of i in the first term of eq. 3.163b requires the lagrangian to be different from the conventional Dirac lagrangian in order for the lagrangian to be real. The simplest, physically acceptable, free spin ½ tachyon lagrangian density for ψ_{CL} is:

$$\mathscr{L}_{CL} = \psi_{CL}{}^{C}(x)(\gamma\cdot\nabla - m)\psi_{CL}(x) \tag{3.168}$$

where

$$\psi_{CL}{}^{C}(x) = [\psi_{CL}(x)]^{\dagger}\big|_{\mathbf{x_i} = -\mathbf{x_i}}\, i\gamma^0\gamma^5 \tag{3.169}$$

similar to eq. 3.128. In words, eq. 3.169 states take the hermitean conjugate of $\psi_{CL}(x)$; change $\mathbf{x_i}$ to $-\mathbf{x_i}$; and then post-multiply by the indicated factors.

The free complexon invariant action (under real Lorentz transformations) is

$$I = \int d^7x \mathscr{L}_{CL} \tag{3.170}$$

The action can be shown to be real

$$I^* = I \tag{3.171}$$

in a manner similar to the case considered in Appendix 3-A. The tachyonic complexon's energy-momentum tensor is

---

[65] Such as G. Feinberg, Phys. Rev. **159**, 1089 (1967).

$$\mathfrak{T}_{CL\mu\nu} = -\, g_{\mu\nu}\, \mathfrak{L}_{CL} + \partial\mathfrak{L}_{CL}/\partial(D^{\mu}\psi_{CL})\, D_{\nu}\psi_{CL} \qquad (3.172)$$
$$= i\psi_{CL}{}^{C}\gamma^{0}\gamma^{5}\gamma_{\mu}D_{\nu}\psi_{CL}$$

where

$$D_0 = \partial/\partial x^0$$
$$D_k = \partial/\partial x^k + i\, \partial/\partial x_i{}^k \qquad (3.129)$$

and thus the conserved energy and momentum are

$$P^0 = H = \int d^3x_r d^3x_i\, \mathfrak{T}_{CL}{}^{00} = i\int d^3x_r d^3x_i \psi_{CL}{}^{C}\gamma^5(\boldsymbol{\alpha}\cdot\mathbf{D} + \beta m)\psi_{CL} \qquad (3.173)$$

and

$$P^k = \int d^3x_r d^3x_i\, \mathfrak{T}_{CL}{}^{0k} = -\, i\int d^3x_r d^3x_i\, \psi_{CL}{}^{C}\gamma^5 D^k\psi_{CL} \qquad (3.174)$$

Having defined a suitable tachyon lagrangian we can now proceed to its canonical quantization. The conjugate momentum can be calculated from the lagrangian density eq. 3.168:

$$\pi_{CLa} = \partial\mathfrak{L}_{CL}/\partial\dot{\psi}_{CLa} \equiv \partial\mathfrak{L}_{CL}/\partial(\partial\psi_{CLa}/\partial t) = -i([\psi_{CL}(x)]^{\dagger}|_{\mathbf{x}_i = -\mathbf{x}_i}\gamma^5)_a \qquad (3.175)$$

The resulting non-zero, canonical anti-commutation relations are presumably

$$\{\pi_{CLa}(x), \psi_{CLb}(y)\} = i\,\delta_{ab}\,\delta^3(x_r - y_r)\delta^3(x_i - y_i)$$

based on locality in both real and imaginary coordinates, or

$$\{\psi_{CL}{}^{\dagger}{}_{a}(x)|_{\mathbf{x}_i = -\mathbf{x}_i}, \psi_{Tb}(y)\} = -\,[\gamma^5]_{ab}\,\delta^3(x_r - y_r)\delta^3(x_i - y_i) \qquad (3.176)$$

At this point we might attempt to complete the canonical quantization procedure in the conventional manner by fourier expanding the field and specifying anti-commutation relations for the fourier component amplitudes. However the incompleteness of the set of plane waves, which are limited by the restriction $\mathbf{p}_r^2 \geq m^2$, causes the equal time anti-commutator of the fields *not* to yield a δ-functions.

Therefore we turn to the previous successful approach to tachyon quantization[66] and decompose the tachyonic complexon field into left-handed and right-handed parts and then second quantize in light-front coordinates.

Separation into Left-Handed and Right-Handed Fields

As before we will use a transformed set of Dirac matrices (eq. 3.32) to develop our left-handed and right-handed tachyon formulations. The γ^5 chirality operator's eigenvalues define handedness: +1 corresponds to right-handed; and −1 corresponds to left-handed:

$$\gamma^5\psi_{CLL} = -\,\psi_{CLL} \qquad\qquad \gamma^5\psi_{CLR} = \psi_{CLR} \qquad (3.177)$$

As before we define left-handed and right-handed tachyon fields with the projection operators:

$$\begin{aligned} C^{\pm} &= \tfrac{1}{2}(I \pm \gamma^5) \\ C^{+} + C^{-} &= I \\ C^{\pm\,2} &= C^{\pm} \\ C^{+}C^{-} &= 0 \end{aligned} \qquad (3.34)$$

with the result

$$\begin{aligned} \psi_{CLL} &= C^{-}\psi_{CL} \\ \psi_{CLR} &= C^{+}\psi_{CL} \end{aligned} \qquad (3.178)$$

[66] Blaha (2006) discusses this case in detail.

We can calculate the commutation relations of the left-handed and right-handed tachyonic complexon fields from eq. 3.176 by pre-multiplying and post-multiplying by $\frac{1}{2}(1 - \gamma^5)$ and $\frac{1}{2}(1 + \gamma^5)$. The results are:

$$\{\psi_{CLLa}^{\dagger}(x)|_{\mathbf{x_i} = -\mathbf{x_i}}, \psi_{CLLb}(y)\} = C^{-}_{ab}\,\delta^6(x - y) \tag{3.179}$$

$$\{\psi_{CLRa}^{\dagger}(x)|_{\mathbf{x_i} = -\mathbf{x_i}}, \psi_{CLRb}(y)\} = -C^{+}_{ab}\,\delta^6(x - y) \tag{3.180}$$

$$\{\psi_{CLLa}^{\dagger}(x)|_{\mathbf{x_i} = -\mathbf{x_i}}, \psi_{CLRb}(y)\} = \{\psi_{CLRa}^{\dagger}(x)|_{\mathbf{x_i} = -\mathbf{x_i}}, \psi_{CLLb}(x')\} = 0 \tag{3.181}$$

where

$$\delta^6(x - y) = \delta^3(x_r - y_r)\delta^3(x_i - y_i) \tag{3.182}$$

The lagrangian density of eq. 3.168 decomposes into left-handed and right-handed parts: (The change $\mathbf{x_i}$ to $-\mathbf{x_i}$ will be understood in $\psi_{CLL}^{\dagger}(x)$ and $\psi_{CLR}^{\dagger}(x)$ in the following.)

$$\mathcal{L}_{CL} = \psi_{CLL}^{\dagger}\gamma^0 i\gamma^\mu\partial_\mu\psi_{CLL} - \psi_{CLR}^{\dagger}\gamma^0 i\gamma^\mu\partial_\mu\psi_{CLR} - im[\psi_{CLR}^{\dagger}\gamma^0\psi_{CLL} - \psi_{CLL}^{\dagger}\gamma^0\psi_{CLR}] \tag{3.183}$$

Further Separation into + and – Light-Front Complexon Fields

As previously we now use light-front coordinates and quantization to obtain a successful second quantization of this form of tachyon field. Light-front variables, in the present case where we have to contend with complex 3-vectors, are defined by the real coordinates and derivatives:

$$x^{\pm} = (x^0 \pm x_r^3)/\surd 2$$

$$\partial/\partial x^{\pm} \equiv \partial^{\mp} \equiv (\partial/\partial x^0 \pm \partial/\partial x_r^3)/\surd 2 \tag{3.184}$$

with the "transverse" real coordinate variables, x_r^1 and x_r^2, and imaginary coordinate variables x_i^1, x_i^2, and x_i^3.

The inner product of two 4-vectors has the form

$$x{\cdot}y = x^+y^- + y^+x^- + i[y_i^3(x^+ - x^-) + x_i^3(y^+ - y^-)]/\surd 2 + x_i^3y_i^3 - (\mathbf{x}_{r\perp} - i\mathbf{x}_{i\perp})\cdot(\mathbf{y}_{r\perp} - i\mathbf{y}_{i\perp}) \tag{3.185}$$

with

$$\mathbf{x}_{r\perp} = (x_r^1, x_r^2) \qquad \mathbf{x}_{i\perp} = (x_i^1, x_i^2) \tag{3.186}$$
$$\mathbf{y}_{r\perp} = (y_r^1, y_r^2) \qquad \mathbf{y}_{i\perp} = (y_i^1, y_i^2)$$

where $x = (x^0, \mathbf{x} = \mathbf{x_r} - i\mathbf{x_i})$ and $y = (y^0, \mathbf{y} = \mathbf{y_r} - i\mathbf{y_i})$. Momenta are always defined as $p = (p^0, \mathbf{p} = \mathbf{p_r} + i\mathbf{p_i})$.

The light-front definition of Dirac matrices is:

$$\gamma^{\pm} = (\gamma^0 \pm \gamma^3)/\surd 2 \tag{3.42}$$

with transverse matrices γ^1 and γ^2 defined as usual. Note:

$$\gamma^{\pm\,2} = 0$$

We define "+" and "–" tachyon fields with the projection operators:

$$R^{\pm} = ½(I \pm \gamma^0\gamma^3) \tag{3.43}$$

Left-handed, ± light-front fields: $\psi_{CLL}{}^{\pm} = R^{\pm}C^{-}\psi_{CL}$ (3.187)

Right-handed, ± light-front fields: $\psi_{CLR}{}^{\pm} = R^{\pm}C^{+}\psi_{CL}$

Transforming to light-front variables and fields as above we obtain the light-front free tachyon lagrangian:

$$\begin{aligned}\mathcal{L}_{CL} = {} & 2^{1/2}\psi_{CLL}^{+\dagger} i\partial^- \psi_{CLL}^{+} + 2^{1/2}\psi_{CLL}^{-\dagger} i\partial^+ \psi_{CLL}^{-} - \psi_{CLL}^{+\dagger}\gamma^0[i\gamma_\perp\cdot\nabla_{r\perp} - \gamma\cdot\nabla_i]\psi_{CLL}^{-} - \\ & - \psi_{CLL}^{-\dagger}\gamma^0[i\gamma_\perp\cdot\nabla_{r\perp} - \gamma\cdot\nabla_i]\psi_{CLL}^{+} - 2^{1/2}\psi_{CLR}^{+\dagger} i\partial^- \psi_{CLR}^{+} - 2^{1/2}\psi_{CLR}^{-\dagger} i\partial^+ \psi_{CLR}^{-} + \\ & + \psi_{CLR}^{+\dagger}\gamma^0[i\gamma_\perp\cdot\nabla_{r\perp} - \gamma\cdot\nabla_i]\psi_{CLR}^{-} + \psi_{CLR}^{-\dagger}\gamma^0[i\gamma_\perp\cdot\nabla_{r\perp} - \gamma\cdot\nabla_i]\psi_{CLR}^{+} - \\ & - im[\psi_{CLR}^{+\dagger}\gamma^0\psi_{CLL}^{-} - \psi_{CLL}^{+\dagger}\gamma^0\psi_{CLR}^{-} + \psi_{CLR}^{-\dagger}\gamma^0\psi_{CLL}^{+} - \psi_{CLL}^{-\dagger}\gamma^0\psi_{CLR}^{+}]\end{aligned} \tag{3.188}$$

(Note the similarity to eq. 3.45 in the previous tachyon case.) Again the difference in signs between the left-handed and right-handed terms will be a crucial factor in the derivation of the left-handed features of the Standard Model.

Eq. 3.188 generates the equations of motion:

$$\begin{aligned} 2^{1/2} i\partial^- \psi_{CLL}^{+} - \gamma^0[i\gamma_\perp\cdot\nabla_{r\perp} - \gamma\cdot\nabla_i]\psi_{CLL}^{-} + im\gamma^0\psi_{CLR}^{-} &= 0 \\ 2^{1/2} i\partial^- \psi_{CLR}^{+} - \gamma^0[i\gamma_\perp\cdot\nabla_{r\perp} - \gamma\cdot\nabla_i]\psi_{CLR}^{-} + im\gamma^0\psi_{CLL}^{-} &= 0 \\ 2^{1/2} i\partial^+ \psi_{CLL}^{-} - \gamma^0[i\gamma_\perp\cdot\nabla_{r\perp} - \gamma\cdot\nabla_i]\psi_{CLL}^{+} + im\gamma^0\psi_{CLR}^{+} &= 0 \\ 2^{1/2} i\partial^+ \psi_{CLR}^{-} - \gamma^0[i\gamma_\perp\cdot\nabla_{r\perp} - \gamma\cdot\nabla_i]\psi_{CLR}^{+} + im\gamma^0\psi_{CLL}^{+} &= 0 \end{aligned} \tag{3.189}$$

Eqs. 3.189 show that ψ_{CLL}^- and ψ_{CLR}^- are dependent fields that are functions of ψ_{CLL}^+ and ψ_{CLR}^+ on the light-front where x^+ equals a constant. They can be expressed in an integral form as well. (The independent fields ψ_{CLL}^+ and ψ_{CLR}^+ play a fundamental role in tachyonic complexon theory and are used to define "in" and "out" tachyon states in perturbation theory.)

The conjugate momenta implied by eq. 3.188 are

$$\begin{aligned} \pi_{CLL}^{+} &= \partial\mathcal{L}/\partial(\partial^-\psi_{CLL}^{+}) = 2^{1/2} i\psi_{CLL}^{+\dagger} \\ \pi_{CLL}^{-} &= \partial\mathcal{L}/\partial(\partial^-\psi_{CLL}^{-}) = 0 \end{aligned} \tag{3.190}$$

$$\begin{aligned} \pi_{CLR}^{+} &= \partial\mathcal{L}/\partial(\partial^-\psi_{CLR}^{+}) = -2^{1/2} i\psi_{CLR}^{+\dagger} \\ \pi_{CLR}^{-} &= \partial\mathcal{L}/\partial(\partial^-\psi_{CLR}^{-}) = 0 \end{aligned} \tag{3.191}$$

x^+ plays the role of the "time" variable in light-front quantized theories. So we define canonical equal x^+ anti-commutation relations for spin ½ tachyonic complexons also.

The canonical equal-light-front ($x^+ = y^+$) anti-commutation relations of the independent fields would normally be:

$$\{\psi_{CLL}{}^{+\dagger}{}_a(x), \psi_{CLL}{}^{+}{}_b(y)\} = 2^{-1}[C^-R^+]_{ab}\delta(x^- - y^-)\delta^2(x_r - y_r)\delta^3(x_I - y_i) \quad (3.192)$$

$$\{\psi_{CLR}{}^{+\dagger}{}_a(x), \psi_{CLR}{}^{+}{}_b(y)\} = -2^{-1}[C^+R^+]_{ab}\,\delta(x^- - y^-)\delta^2(x_r - y_r)\delta^3(x_I - y_i) \quad (3.193)$$

$$\{\psi_{CLL}{}^{+\dagger}{}_a(x), \psi_{CLR}{}^{+}{}_b(y)\} = \{\psi_{CLR}{}^{+\dagger}{}_a(x), \psi_{CLL}{}^{+}{}_b(y)\} = 0 \quad (3.194)$$

$$\{\psi_{CLL}{}^{+}{}_a(x), \psi_{CLR}{}^{+}{}_b(y)\} = \{\psi_{CLR}{}^{+\dagger}{}_a(x), \psi_{CLL}{}^{+\dagger}{}_b(y)\} = 0 \quad (3.195)$$

But as in the previous case they will be modified.

Again we see that the right-handed tachyon anti-commutation relation (eq. 3.193) has a minus sign relative to the corresponding conventional right-handed anti-commutation relation (eq. 3.55).

The sign differences between the left-handed and right-handed lagrangian terms ultimately lead to parity violating features in the Standard Model lagrangian.

Left-Handed Tachyonic Complexons

The free, "+" light-front, left-handed tachyonic complexon Fourier expansion is:

$$\psi_{CLL}{}^{+}(x_r, x_i) = \sum_{\pm s} \int d^2p_r dp^+ d^3p_i\, N_{CLL}{}^{+}(p)\theta(p^+)\delta((p_i{}^3(p^+ - p^-)/\sqrt{2} + \mathbf{p}_{r\perp}\cdot\mathbf{p}_{i\perp})/m^2)\cdot$$

$$\cdot[b_{CLL}{}^{+}(p, s)u_{CLL}{}^{+}(p, s)e^{-i(p.x + p^*.x^*)/2} + d_{CLL}{}^{+\dagger}(p, s)v_{CLL}{}^{+}(p, s)e^{+i(p.x + p^*.x^*)/2}] \quad (3.196)$$

(Compare to eq. 3.109.) Its hermitean conjugate is

$$\psi_{CLL}^{+\dagger}(x_r, x_i) = \sum_{\pm s}\int d^2p_r dp^+ d^3p_i\, N_{CLL}^{+}(p)\theta(p^+)\delta((p_i^3(p^+ - p^-)/\surd 2 + \mathbf{p}_{r\perp}\cdot\mathbf{p}_{i\perp})/m^2)\cdot$$
$$\cdot[b_{CLL}^{\dagger}(p^*,s)u_{CLL}^{\dagger}(p^*,s)e^{+i(p^*.x + p.x^*)/2} + d_{CLL}(p^*,s)v_{CLL}^{\dagger}(p^*,s)e^{-i(p^*.x + p.x^*)/2}] \quad (3.197)$$

where $\mathbf{p} = \mathbf{p_r} + i\mathbf{p_i}$ (eq. 3.95), $\mathbf{x} = \mathbf{x_r} - i\mathbf{x_i}$, $p\cdot x = p^0x^0 - \mathbf{p\cdot x}$, and † indicates hermitean conjugate. (Compare to eq. 3.110.) The spinors are

$$\begin{aligned}
u_{CLL}^{+}(p, s) &= C^- R^+ S_{CL}w^1(0)\\
u_{CLL}^{+}(p, -s) &= C^- R^+ S_{CL}w^2(0)\\
v_{CLL}^{+}(p, s) &= C^- R^+ S_{CL}w^3(0)\\
v_{CLL}^{+}(p, -s) &= C^- R^+ S_{CL}w^4(0)\\
u_{CLL}^{+\dagger}(p^*, s) &= w^{1T}(0)S_{CL}R^+C^-\\
u_{CLL}^{+\dagger}(p^*, -s) &= w^{2T}(0)S_{CL}R^+C^-\\
v_{CLL}^{+\dagger}(p^*, s) &= w^{3T}(0)S_{CL}R^+C^-\\
v_{CLL}^{+\dagger}(p^*, -s) &= w^{4T}(0)S_{CL}R^+C^-
\end{aligned} \quad (3.198)$$

using eq. 3.166 where the superscript "T" indicates the transpose (These spinors are described in Appendix 3-A.) and

$$N_{CLL}^{+}(p) = (2\pi)^{-3}(2m/p^+)^{1/2} \quad (3.199)$$

The anti-commutation relations of the Fourier coefficient operators are

$$\{b_{CLL}(p,s), b_{CLL}^{\dagger}(p'^*,s')\} = 2^{-1/2}\,\delta_{ss'}\delta(p^+ - p'^+)\delta^2(\mathbf{p_r} - \mathbf{p'_{r'}})\delta^3(\mathbf{p_i} + \mathbf{p'_{i'}})$$

$$\{d_{CLL}(p,s),\ d_{CLL}^{\dagger}(p'^*,s')\} = 2^{-1/2}\delta_{ss'}\delta(p^{+} - p'^{+})\delta^2(\mathbf{p_r} - \mathbf{p'_{r'}})\delta^3(\mathbf{p_i} + \mathbf{p'_{i'}})$$
$$\{b_{CLL}(p,s),\ b_{CLL}(p'^*,s')\} = \{d_{CLL}(p,s),\ d_{CLL}(p'^*,s')\} = 0$$
$$\{b_{CLL}^{\dagger}(p,s),\ b_{CLL}^{\dagger}(p'^*,s')\} = \{d_{CLL}^{\dagger}(p,s),\ d_{CLL}^{\dagger}(p'^*,s')\} = 0 \qquad (3.200)$$
$$\{b_{CLL}(p,s),\ d_{CLL}^{\dagger}(p'^*,s')\} = \{d_{CLL}(p,s),\ b_{CLL}^{\dagger}(p'^*,s')\} = 0$$
$$\{b_{CLL}^{\dagger}(p,s),\ d_{CLL}^{\dagger}(p'^*,s')\} = \{d_{CLL}(p,s),\ b_{CLL}(p'^*,s')\} = 0$$

The delta-function arguments $\delta^3(\mathbf{p_i} + \mathbf{p'_{i'}})$ above have a positive sign as in eq. 3.138 in order to obtain $\delta^3(\mathbf{x_i} - \mathbf{y_i})$ in the field anti-commutators.

The spinors, eq. 3.198, satisfy

$$\sum_{\pm s} u_{CLL}{}^{+}{}_{\alpha}(p,\ s)\bar{u}_{CLL}{}^{+}{}_{\beta}(p^*,\ s) = (2m)^{-1}[C^{-}R^{+}(i\not{p} + m)R^{-}C^{+}]_{\alpha\beta} \qquad (3.201)$$
$$\sum_{\pm s} v_{CLL}{}^{+}{}_{\alpha}(p,\ s)\bar{v}_{CLL}{}^{+}{}_{\beta}(p^*,\ s) = (2m)^{-1}[C^{-}R^{+}(i\not{p} - m)R^{-}C^{+}]_{\alpha\beta}$$

where $\bar{u}_{CLL}{}^{+} = u_{CLL}{}^{+\dagger}\gamma^0$ and $\bar{v}_{CLL}{}^{+} = v_{CLL}{}^{+\dagger}\gamma^0$.

We now evaluate the canonical left-handed, light-front anti-commutation relation (eq. 3.192):

$$\{\psi_{CLL}{}^{+}{}_{a}(x),\ \psi_{CLL}{}^{+\dagger}{}_{b}(y)\} = \sum_{\pm s,s'} \int d^3p_i d^2p dp^{+} \int d^3p_i' d^2p' dp'^{+} N_{CLL}{}^{+}(p)\, N_{CLL}{}^{+}(p')\cdot$$
$$\cdot\theta(p^{+})\theta(p'^{+})\delta((p_i{}^3(p^{+}-p^{-})/\surd 2 + \mathbf{p_{r\perp}}\cdot\mathbf{p_{i\perp}})/m^2)\, \delta((p_i'^3(p'^{+} - p'^{-})/\surd 2 + \mathbf{p'_{r\perp}}\cdot\mathbf{p'_{i\perp}})/m^2)\cdot$$
$$\cdot[\{b_{CLL}{}^{+\dagger}(p'^*,s'),b_{CLL}{}^{+}(p,s)\}u_{CLL}{}^{+}{}_{a}(p,s)u_{CLL}{}^{+\dagger}{}_{b}(p'^*,s')e^{+i(p'^*\cdot y+p'\cdot y^*)2 - i(p\cdot x+p^*\cdot x^*)/2} +$$
$$+\{d_{CLL}{}^{+}(p'^*,s'),d_{CLL}{}^{+\dagger}(p,s)\}v_{CLL}{}^{+}{}_{a}(p,s)v_{CLL}{}^{+\dagger}{}_{b}(p'^*,s')e^{-i(p'^*\cdot y+p'\cdot y^*)/2 + i(p\cdot x + p^*\cdot x^*)/2}]$$
$$= 2^{-1/2}\sum_{\pm s} \int d^3p_i d^2p_r dp^{+}\, [N_{CLL}{}^{+}(p)]^2\theta(p^{+})[\delta((p_i{}^3(p^{+} - p^{-})/\surd 2 + \mathbf{p_{r\perp}}\cdot\mathbf{p_{i\perp}})/m^2)]^2$$
$$[u_{CLL}{}^{+}{}_{a}(p,s)u_{CLL}{}^{+}{}_{b}{}^{\dagger}(p^*,s)e^{+i(p^*\cdot(y-x)+p\cdot(y^*-x^*))/2} +$$

$$+ v_{CLL}{}^{+}{}_{a}(p,s)v_{CLL}{}^{+\dagger}{}_{b}(p^*,s)e^{-i(p^*\cdot(y-x)+p\cdot(y^*-x^*))/2}]$$

and, using eq. 3.141,

$$= -2^{-3/2}\int d^3p_i d^2p dp^+\theta(p^+)[N_{CLL}{}^{+}(p)]^2\delta'((p_i{}^3(p^+ - p^-)/\surd 2 + \mathbf{p}_{r\perp}\cdot\mathbf{p}_{i\perp})/m^2)(2m)^{-1}$$
$$\{[C^-R^+(i\not{p} + m)\gamma^0R^+C^-]_{ab}e^{+i(p^*\cdot(y-x)+p\cdot(y^*-x^*))/2} +$$
$$+ [C^-R^+(i\not{p} - m)\gamma^0R^+C^-]_{ab}e^{-i(p^*\cdot(y-x)+p\cdot(y^*-x^*))/2}\}$$

$$= -(1/2)C^-R^+\delta_{ab}\int d^3p_i\, d^2p_\perp\int_0^\infty dp^+\,\delta'((p_i{}^3(p^+ - p^-)/\surd 2 + \mathbf{p}_\perp\cdot\mathbf{p}_{i\perp})/m^2)(2\pi)^{-6}\cdot$$
$$\cdot\{e^{+i\{p^+(y^- - x^-) - \mathbf{p}_{r\perp}\cdot(\mathbf{y}_{r\perp} - \mathbf{x}_{r\perp}) + \mathbf{p}_i\cdot(\mathbf{y}_i - \mathbf{x}_i)\}} + e^{-i\{p^+(y^- - x^-) - \mathbf{p}_{r\perp}\cdot(\mathbf{y}_{r\perp} - \mathbf{x}_{r\perp}) + \mathbf{p}_i\cdot(\mathbf{y}_i - \mathbf{x}_i)\}}\}$$

$$= -C^-R^+\delta_{ab}(4\pi)^{-1}\int_0^\infty dp^+\delta'(\nabla_r\cdot\nabla_i/m^2)\delta^3(y_i - x_i)\,\delta^2(y_r - x_r)\{e^{+ip^+(y^- - x^-)} + e^{-ip^+(y^- - x^-)}\}$$

whereupon we revert back to the original form of the constraint: $\delta(\nabla_r\cdot\nabla_i/m^2)$

$$\{\psi_{CLL}{}^{+}{}_{a}(x), \psi_{CLL}{}^{+\dagger}{}_{b}(y)\} = -(1/2)C^-R^+\delta_{ab}\,\delta'(\nabla_r\cdot\nabla_i/m^2)\delta(y^- - x^-)\delta^2(y_r - x_r)\delta^3(y_i - x_i)$$
$$(3.202)$$

The result is the left-handed, light-front equivalent of the earlier non-tachyon result eq. 3.142. Again the constraint is apparent in the anti-commutator. (The factor of 2 difference is due to light-front coordinate definitions.)

Therefore we have left-handed, light-front quantized tachyonic complexons with the equivalent of canonical anti-commutation relations, and with localized tachyonic complexons. As a result we have a canonical tachyonic complexon Quantum Field Theory

Left-handed Case 4 Tachyonic Complexon Feynman Propagator

The light-front Feynman propagator for the left-handed $\psi_{CLL}^{\ +}$ *tachyonic* complexon field is

$$iS^{+}_{CLLF}(x,y) = \theta(x^{+} - y^{+})\langle 0|\psi_{CLL}^{\ +}(x)\psi_{CLL}^{\ +\dagger}(y)\gamma^{0}|0\rangle - \\ - \theta(y^{+} - x^{+})\langle 0|\psi_{CLL}^{\ +\dagger}(y)\gamma^{0}\psi_{CLL}^{\ +}(x)|0\rangle \tag{3.203}$$

$$= -\tfrac{1}{2}\int d^3p_i d^2p_r dp^{+}\theta(p^{+})N_{CLL}^{\ +2}\delta'((p_i^{\ 3}(p^{+} - p^{-})/\sqrt{2} + \mathbf{p}_{r\perp}\cdot\mathbf{p}_{i\perp})/m^2)(2m)^{-1}C^{-}R^{+} \\ \{\theta(x^{+} - y^{+})[(i\not{p} + m)\gamma^{0}]e^{+i(p^{*}\cdot(y-x)+p\cdot(y^{*}-x^{*}))/2} + \\ + \theta(y^{+} - x^{+})[(i\not{p} - m)\gamma^{0}]e^{-i(p^{*}\cdot(y-x)+p\cdot(y^{*}-x^{*}))/2}\}R^{+}C^{-}\gamma^{0}$$

If we define the on-shell momentum variable $p_0^{\ -} = (p_{r0}^{\ 1}p_{r0}^{\ 1} + p_{r0}^{\ 2}p_{r0}^{\ 2} - \mathbf{p}_{i0}\cdot\mathbf{p}_{i0} - m^2)/(2p_0^{\ +})$, $p_0^{\ +} = p^{+}$, $p_{r0}^{\ j} = p_r^{\ j}$ (for j = 1, 2), $\mathbf{p}_{i0} = \mathbf{p}_i$, $p_{r\perp 0}^{\ 2} = p_{r0}^{\ j}p_{r0}^{\ j}$ and $\not{p}_0 = p_0\cdot\gamma$ with $p_0 = (p^0, \mathbf{p}_{r0} + i\mathbf{p}_{r0})$ then the above equation can be rewritten as

$$= -\tfrac{1}{2}C^{-}R^{+}\int d^4p d^3p_i N_{CLL}^{\ +2}\delta'((p_{i0}^{\ 3}(p_0^{\ +} - p_0^{\ -})/\sqrt{2} + \mathbf{p}_{r\perp 0}\cdot\mathbf{p}_{i\perp 0})/m^2)(4\pi m)^{-1}e^{+i(p^{*}\cdot(y-x)+p\cdot(y^{*}-x^{*}))/2}$$

$$\cdot\{\theta(p^{+})(i\not{p}_0 + m)\gamma^{0}]/[p^{-} - p_0^{\ -} + i\varepsilon] + \theta(-p^{+})(i\not{p}_0 - m)\gamma^{0}]/[p^{-} + p_0^{\ -} - i\varepsilon]\}R^{+}C\gamma^{0}$$

$$= -\tfrac{1}{2}\int d^4p_r d^3p_i\, N_{CLL}^{\ +2}\delta'((p_{i0}^{\ 3}(p^{+} - p^{-})/\sqrt{2} + \mathbf{p}_{r\perp}\cdot\mathbf{p}_{i\perp})/m^2)(p^{+}/4\pi m)\, e^{+i(p^{*}\cdot(y-x)+p\cdot(y^{*}-x^{*}))/2}\,\cdot \\ \cdot[C^{-}R^{+}(i\not{p} + m)\gamma^{0}R^{+}C^{-}\gamma^{0}][(p^2 + m^2 + i\varepsilon)]^{-1}$$

with $p_r = (p^0, \mathbf{p}_r)$ and $p = (p^0, \mathbf{p}_r + i\mathbf{p}_r)$. Substituting for N_{CLL} and using $x\delta'(x) = -\delta(x)$ we obtain

$$= -\tfrac{1}{2}\int d^4p_r d^3p_i (2\pi)^{-7}\delta'(\mathbf{p_r}\cdot\mathbf{p_i}/m^2)\exp[ip^0(y^0 - x^0) - i\mathbf{p_r}\cdot(\mathbf{y_r} - \mathbf{x_r}) + i\mathbf{p_i}\cdot(\mathbf{y_i} - \mathbf{x_i})]\cdot$$
$$\cdot\,[C^-R^+(i\not{p} + m)R^-C^+]/(p^2 + m^2 + i\varepsilon)$$

since $C^-R^+(i\not{p} + m)\gamma^0R^+C^-\gamma^0 = C^-R^+(i\not{p} + m)R^-C^+$. The integral can be written:

$$I = \int d^4p_r d^3p_i \delta'(\mathbf{p_r}\cdot\mathbf{p_i}/m^2)C^-R^+(i\not{p} + m)R^-C^+\cdot$$
$$\cdot\exp[-ip^0(x^0 - y^0) + i\mathbf{p_r}\cdot(\mathbf{x_r} - \mathbf{y_r}) - i\mathbf{p_i}\cdot(\mathbf{x_i} - \mathbf{y_i})]/(p^2 + m^2 + i\varepsilon)$$
$$= \int d^4p_r dM^2\delta'(\nabla_r\cdot\nabla_i/m^2)C^-R^+(ip^0\gamma^0 - (\nabla_r - i\nabla_i)\cdot\gamma + m)R^-C^+\cdot$$
$$\cdot\exp[-ip^0(x^0 - y^0) + i\mathbf{p_r}\cdot(\mathbf{x_r} - \mathbf{y_r})]J_2(\mathbf{x_i} - \mathbf{y_i}, M^2)/(p_r^2 + M^2 + i\varepsilon)$$

where

$$J_2(\mathbf{x_i} - \mathbf{y_i}, M^2) = (2\pi)^{-3}\int d^3p_i\, \delta(M^2 - \mathbf{p_i}^2 - m^2)\exp[-i\mathbf{p_i}\cdot(\mathbf{x_i} - \mathbf{y_i})] \quad (3.204)$$

$$= (2\pi)^{-2}|\mathbf{x_i} - \mathbf{y_i}|^{-1}\theta(M^2 - m^2)\sin((M^2 - m^2)^{1/2}|\mathbf{x_i} - \mathbf{y_i}|)$$

This tachyonic complexon Feynman propagator can be rearranged into the form of a spectral integral:

$$iS^+_{CLLF}(x, y) = -\int dM\, C^-R^+(\gamma^0\partial/\partial x^0 + (\nabla_r - i\nabla_i)\cdot\gamma - m)R^-C^+\,\delta'(\nabla_r\cdot\nabla_i/m^2)\cdot$$
$$\cdot J_2(\mathbf{x_i} - \mathbf{y_i}, M^2)\Delta_{FT}(x - y, M) \quad (3.205)$$

with ∇_r and ∇_i derivatives with respect to $\mathbf{x_r}$ and $\mathbf{x_i}$ and where

$$\Delta_{FT}(x - y, M) = (2\pi)^{-4}\int d^4p_r \exp[-ip^0(x^0 - y^0) + i\mathbf{p_r}\cdot(\mathbf{x_r} - \mathbf{y_r})]/(p_r^2 + M^2 + i\varepsilon)$$
$$(3.206)$$

Right-Handed Tachyonic Complexons

The case of right-handed tachyonic complexons is similar to left-handed complexons with only one difference: a minus sign in the canonical right-handed equal-time commutation relations resulting in a minus sign in the creation and annihilation operator anti-commutation relations. The right-handed tachyonic complexon wave function (eq. 3.187) light-front Fourier expansion is:

$$\overset{+}{\psi}_{CLR}(x_r, x_i) = \sum_{\pm s} \int d^2p_r dp^+ d^3p_i \, \overset{+}{N}_{CLR}(p)\theta(p^+)\delta((p_i^3(p^+ - p^-)/\surd 2 + \mathbf{p}_{r\perp}\cdot\mathbf{p}_{i\perp})/m^2)\cdot$$
$$\cdot[\overset{+}{b}_{CLR}(p, s)\overset{+}{u}_{CLR}(p, s)e^{-i(p.x + p^*.x^*)/2} + \overset{+}{d}{}^{\dagger}_{CLR}(p, s)\overset{+}{v}_{CLR}(p, s)e^{+i(p.x + p^*.x^*)/2}] \tag{3.207}$$

where

$$\overset{+}{N}_{CLR}(p) = (2\pi)^{-3}(2m/p^+)^{1/2} \tag{3.208}$$

Its hermitean conjugate is

$$\overset{+}{\psi}{}^{\dagger}_{CLR}(x_r, x_i) = \sum_{\pm s} \int d^2p_r dp^+ d^3p_i \, \overset{+}{N}_{CLR}(p)\theta(p^+)\delta((p_i^3(p^+ - p^-)/\surd 2 + \mathbf{p}_{r\perp}\cdot\mathbf{p}_{i\perp})/m^2)\cdot$$
$$\cdot[b^{\dagger}_{CLR}(p^*,s)u^{\dagger}_{CLR}(p^*,s)e^{+i(p^*.x + p.x^*)/2} + d_{CLR}(p^*,s)v^{\dagger}_{CLR}(p^*,s)e^{-i(p^*.x + p.x^*)/2}] \tag{3.209}$$

where $\mathbf{p} = \mathbf{p_r} + i\mathbf{p_i}$ (eq. 3.95), $\mathbf{x} = \mathbf{x_r} - i\mathbf{x_i}$, $p\cdot x = p^0x^0 - \mathbf{p}\cdot\mathbf{x}$, and † indicates hermitean conjugate. (Compare to eq. 3.110.) The right-handed spinors are

$$\overset{+}{u}_{CLR}(p, s) = C^+ \overset{+}{R} S_{CR} w^1(0)$$
$$\overset{+}{u}_{CLR}(p, -s) = C^+ \overset{+}{R} S_{CR} w^2(0)$$
$$\overset{+}{v}_{CLR}(p, s) = C^+ \overset{+}{R} S_{CR} w^3(0)$$
$$\overset{+}{v}_{CLR}(p, -s) = C^+ \overset{+}{R} S_{CR} w^4(0) \tag{3.210}$$

$$u_{CLR}^{+\dagger}(p^*, s) = w^{1T}(0)S_{CR}R^{+}C^{+}$$
$$u_{CLR}^{+\dagger}(p^*, -s) = w^{2T}(0)S_{CR}R^{+}C^{+}$$
$$v_{CLR}^{+\dagger}(p^*, s) = w^{3T}(0)S_{CR}R^{+}C^{+}$$
$$v_{CLR}^{+\dagger}(p^*, -s) = w^{4T}(0)S_{CR}R^{+}C^{+}$$

where the superscript "T" indicates the transpose. The anti-commutation relations of the Fourier coefficient operators are

$$\begin{aligned}
\{b_{CLR}(p,s), b_{CLR}^{\dagger}(p'^*,s')\} &= -2^{-1/2}\delta_{ss'}\delta(p^{+} - p'^{+})\delta^2(\mathbf{p_r} - \mathbf{p'_{r'}})\delta^3(\mathbf{p_i} + \mathbf{p'_{i'}}) \\
\{d_{CLR}(p,s), d_{CLR}^{\dagger}(p'^*,s')\} &= -2^{-1/2}\delta_{ss'}\delta(p^{+} - p'^{+})\delta^2(\mathbf{p_r} - \mathbf{p'_{r'}})\delta^3(\mathbf{p_i} + \mathbf{p'_{i'}}) \\
\{b_{CLR}(p,s), b_{CLR}(p'^*,s')\} &= \{d_{CLR}(p,s), d_{CLR}(p'^*,s')\} = 0 \\
\{b_{CLR}^{\dagger}(p,s), b_{CLR}^{\dagger}(p'^*,s')\} &= \{d_{CLR}^{\dagger}(p,s), d_{CLR}^{\dagger}(p'^*,s')\} = 0 \qquad (3.211) \\
\{b_{CLR}(p,s), d_{CLR}^{\dagger}(p'^*,s')\} &= \{d_{CLR}(p,s), b_{CLR}^{\dagger}(p'^*,s')\} = 0 \\
\{b_{CLR}^{\dagger}(p,s), d_{CLR}^{\dagger}(p'^*,s')\} &= \{d_{CLR}(p,s), b_{CRR}(p'^*,s')\} = 0
\end{aligned}$$

The spinors satisfy

$$\sum_{\pm s} u_{CLR}^{+}{}_{\alpha}(p, s)\bar{u}_{CLR}^{+}{}_{\beta}(p^*, s) = (2m)^{-1}[C^{+}R^{+}(-i\not{p} + m)R^{-}C^{-}]_{\alpha\beta} \qquad (3.212)$$
$$\sum_{\pm s} v_{CLR}^{+}{}_{\alpha}(p, s)\bar{v}_{CLR}^{+}{}_{\beta}(p^*, s) = (2m)^{-1}[C^{+}R^{+}(-i\not{p} - m)R^{-}C^{-}]_{\alpha\beta}$$

where $\bar{u}_{CLR}^{+} = u_{CLR}^{+\dagger}\gamma^0$ and $\bar{v}_{CLR}^{+} = v_{CLR}^{+\dagger}\gamma^0$.

The right-handed anti-commutation relation with a minus sign follows in particular because of the minus signs in eqs. 3.211.

Case 4 Right-handed Tachyonic Complexon Feynman Propagator

The Feynman propagator for right-handed tachyonic complexons can be obtained from eqs. 3.205 and 3.206 by changing the parity projection operator and some numerator signs in the integral (basically p → –p) resulting in

$$iS^{+}_{CLRF}(x, y) = \int dM\, C^{+}R^{+}(\gamma^{0}\partial/\partial x^{0} + (\nabla_r - i\nabla_i)\cdot\gamma - m)R^{-}C^{-}\delta'(\nabla_r\cdot\nabla_i/m^2)\cdot$$
$$\cdot J_2(\mathbf{x}_i - \mathbf{y}_i, M^2)\Delta_{FT}(x - y, M) \qquad (3.213)$$

with $\nabla_r + i\nabla_i$ derivatives with respect to $\mathbf{x}_r$ and $\mathbf{x}_i$ and where

$$\Delta_{FT}(x - y, M) = (2\pi)^{-4}\int d^4p_r \exp[-ip^0(x^0 - y^0) + i\mathbf{p}_r\cdot(\mathbf{x}_r - \mathbf{y}_r)]/(p_r^2 + M^2 + i\varepsilon)$$
$$(3.214)$$

Other Cases? No

The four cases considered above are the only cases having symmetry under the real Lorentz group L and a single real energy (with a corresponding single real time parameter) that is independent of the direction of the boost thus preserving (real) spatial rotation invariance. The realness of the time variable survives the breakdown to conventional Lorentz invariance.

One might think that using the other type of spinor boost operator (Compare to eq. 3.154.)

$$S_{CR}(\Lambda_{CR}(\omega, \hat{\mathbf{w}})) = \exp(-i\omega_R\sigma_{0i}w_i/2) = \exp(-\omega_R\gamma^0\boldsymbol{\gamma}\cdot\hat{\mathbf{w}}/2) \qquad (3.215)$$

$$= \cosh(\omega_R/2)I + \sinh(\omega_R/2)\gamma^0\boldsymbol{\gamma}\cdot\hat{\mathbf{w}}$$

where $\omega_R = \omega - i\pi/2$ might lead to more possible forms of spin ½ wave equations and particles. In fact it merely leads to the same particle types but with the role of the left-handed and right-handed fields reversed. The result would be a "right-handed" Standard Model contrary to experiment.

3.8 L_C Spinor Lorentz Boosts Generate 4 Types of Particles that suggest they are Analogues of Leptons and Color Quarks

In this chapter we have found four types of fermions using a complexified form of Lorentz boosts, L_C boosts, that correspond in a natural way with the four general types of known fermions: charged leptons, neutrinos, up-type color quarks and down-type color quarks.[67]

Charged lepton fermions

The conventional Dirac equation and solutions.

Neutrinos

Simple tachyons with real energy and 3-momentum. Their free field equation is:

$$(\gamma^{\mu}\partial/\partial x^{\mu} - m)\psi_T(x) = 0 \qquad (3.18)$$

and their left-handed $\psi_{TL}{}^{+}$ Feynman propagator is:

$$iS^{+}{}_{TLF}(x, y) = \tfrac{1}{2}C^{-}R^{+}\gamma^{0}\int d^4p(2\pi)^{-4}\, p^{+}e^{-ip\cdot(x-y)}/(p^2 + m^2 + i\varepsilon) \qquad (3.76)$$

Similarly the light-front Feynman propagator for the right-handed $\psi_{TR}{}^{+}$ tachyon field is

$$iS^{+}{}_{TRF}(x,y) = -\tfrac{1}{2}C^{+}R^{+}\gamma^{0}\int d^4p(2\pi)^{-4}\, p^{+}e^{-ip\cdot(x-y)}/(p^2 + m^2 + i\varepsilon) \qquad (3.77)$$

---

[67] We call each type of fermion a *species*. Each species has three known generations.

Up-type Color Quarks

Up-type quarks are assumed[68] to be fermions with complex 3-momenta - complexons, and an internal color SU(3) symmetry, that are "normal" with $p^2 = m^2$. Their field equation with a color SU(3) index, denoted a, inserted is

$$[i\gamma^0\partial/\partial t + i\gamma\cdot(\nabla_r + i\nabla_i) - m]\psi_C{}^a(t, \mathbf{x}_r, \mathbf{x}_i) = 0 \qquad (3.101)$$

with the subsidiary condition

$$\nabla_r\cdot\nabla_i\, \psi_C{}^a(t, \mathbf{x}_r, \mathbf{x}_i) = 0 \qquad (3.102a)$$

The free field solution is:

$$\psi_C{}^a(x) = \sum_{\pm s} \int d^3p_r d^3p_i\, N_C(p)\delta(\mathbf{p}_r\cdot\mathbf{p}_i/m^2)[b_C(p,a,s)u_C{}^a(p, s)e^{-i(p.x + p^*.x^*)/2} + $$
$$+ d_C{}^\dagger(p,a,s)v_C{}^a(p, s)e^{+i(p.x + p^*.x^*)/2}] \qquad (3.125a)$$

The free Feynman propagator arranged into the form of a spectral integral is

$$iS_C{}^{ab}(x,y) = -\delta^{ab}\int dM\, (i\gamma^0\partial/\partial x^0 - i(\nabla_r - i\nabla_i)\cdot\gamma + m)\delta'(\nabla_r\cdot\nabla_i/m^2)\cdot$$
$$\cdot J(\mathbf{x}_i - \mathbf{y}_i, M^2)\Delta_F(x - y, M) \qquad (3.149)$$

where

$$\Delta_F(x - y, M) = (2\pi)^{-4}\int d^4p_r\, \exp[-ip^0(x^0 - y^0) + i\mathbf{p}_r\cdot(\mathbf{x}_r - \mathbf{y}_r)]/(p_r{}^2 - M^2 + i\varepsilon)$$
$$(3.150)$$

---

[68] The complexon theory that we develop and use for quark dynamics in the Standard Model is <u>not</u> required. Our Standard Model could use Dirac fermion dynamics for the up-type quarks and tachyon dynamics for down-type quarks. Then the (broken) Left-handed Extended Lorentz group would be the basic space-time group rather than L_C. We choose to use complexon dynamics for quarks because they have an internal SU(3)-like structure suggestive of color SU(3). More importantly, their spin dynamics is different and thus may resolve the differences between theory and experiment for the deep inelastic parton spin-dependent structure functions.

and

$$J(\mathbf{x_i}, M^2) = (2\pi)^{-3}\int d^3p_i\,\delta(M^2 + \mathbf{p_i}^2 - m^2)\exp[-i\mathbf{p_i}\cdot(\mathbf{x_i} - \mathbf{y_i})] \qquad (3.148)$$

$$= (2\pi)^{-2}|\mathbf{x_i} - \mathbf{y_i}|^{-1}\theta(m^2 - M^2)\sin((m^2 - M^2)^{1/2}|\mathbf{x_i} - \mathbf{y_i}|)$$

Down-type Color Quarks

Tachyonic complexons with complex 3-momenta, and an internal global SU(3) symmetry, that have mass shell condition $p^2 = -m^2$. Their field equation with a color SU(3) index, denoted a, inserted is

$$[\gamma^0\partial/\partial t + \boldsymbol{\gamma}\cdot(\nabla_r + i\nabla_i) - m]\psi_{CL}{}^a(t, \mathbf{x_r}, \mathbf{x_i}) = 0 \qquad (3.163a)$$

with the subsidiary condition on the wave function

$$\nabla_r\cdot\nabla_i\,\psi_{CL}{}^a(t, \mathbf{x_r}, \mathbf{x_i}) = 0 \qquad (3.164)$$

Its free field left-handed solution is:

$$\psi_{CLL}{}^{+a}(x_r, x_i) = \sum_{\pm s}\int d^2p_r dp^+ d^3p_i\, N_{CLL}{}^+(p)\theta(p^+)\delta((p_i^3(p^+ - p^-)/\surd 2 + \mathbf{p}_{r\perp}\cdot\mathbf{p}_{i\perp})/m^2)\cdot$$
$$\cdot[b_{CLL}{}^+(p,a,s)u_{CLL}{}^a(p,a,s)e^{-i(p.x + p^*.x^*)/2} + d_{CLL}{}^{+\dagger}(p,a,s)v_{CLL}{}^{+a}(p,a,s)e^{+i(p.x + p^*.x^*)/2}] \qquad (3.196)$$

and its right-handed solution is

$$\psi_{CLR}{}^{+a}(x_r, x_i) = \sum_{\pm s}\int d^2p_r dp^+ d^3p_i\, N_{CLR}{}^+(p)\theta(p^+)\delta((p_i^3(p^+ - p^-)/\surd 2 + \mathbf{p}_{r\perp}\cdot\mathbf{p}_{i\perp})/m^2)\cdot$$

$$\cdot[b_{CLR}{}^+(p,a,s)u_{CLR}{}^{+a}(p,a,s)e^{-i(p.x+p^*.x^*)/2} + d_{CLR}{}^{+\dagger}(p,a,s)v_{CLR}{}^{+a}(p,a,s)e^{+i(p.x+p^*.x^*)/2}] \qquad (3.207)$$

The free left-handed Feynman propagator arranged into the form of a spectral integral is

$$iS^{+}_{CLLF}{}^{ab}(x,y) = -\delta^{ab}\int dM\, C^{-}R^{+}(\gamma^{0}\partial/\partial x^{0} + (\nabla_{r} - i\nabla_{i})\cdot\gamma - m)R^{-}C^{+}\,\delta'(\nabla_{r}\cdot\nabla_{i}/m^{2})\cdot$$
$$\cdot J_{2}(\mathbf{x_{i}} - \mathbf{y_{i}}, M^{2})\Delta_{FT}(x - y, M) \quad (3.205)$$

with ∇_{r} and ∇_{i} derivatives with respect to $\mathbf{x_{r}}$ and $\mathbf{x_{i}}$ and where

$$\Delta_{FT}(x - y, M) = (2\pi)^{-4}\int d^{4}p_{r}\, \exp[-ip^{0}(x^{0} - y^{0}) + i\mathbf{p_{r}}\cdot(\mathbf{x_{r}} - \mathbf{y_{r}})]/(p_{r}^{2} + M^{2} + i\varepsilon) \quad (3.206)$$

and

$$J_{2}(\mathbf{x_{i}}, M^{2}) = (2\pi)^{-3}\int d^{3}p_{i}\, \delta(M^{2} - \mathbf{p_{i}}^{2} - m^{2})\exp[-i\mathbf{p_{i}}\cdot(\mathbf{x_{i}} - \mathbf{y_{i}})] \quad (3.204)$$

$$= (2\pi)^{-2}|\mathbf{x_{i}} - \mathbf{y_{i}}|^{-1}\theta(M^{2} - m^{2})\sin((M^{2} - m^{2})^{1/2}|\mathbf{x_{i}} - \mathbf{y_{i}}|)$$

The free right-handed Feynman propagator arranged into the form of a spectral integral is

$$iS^{+}_{CLRF}{}^{ab}(x, y) = \delta^{ab}\int dM\, C^{+}R^{+}(\gamma^{0}\partial/\partial x^{0} + (\nabla_{r} - i\nabla_{i})\cdot\gamma - m)R^{-}C^{-}\,\delta'(\nabla_{r}\cdot\nabla_{i}/m^{2})\cdot$$
$$\cdot J_{2}(\mathbf{x_{i}} - \mathbf{y_{i}}, M^{2})\Delta_{FT}(x - y, M) \quad (3.213)$$

with ∇_{r} and ∇_{i} derivatives with respect to $\mathbf{x_{r}}$ and $\mathbf{x_{i}}$, and where

$$\Delta_{FT}(x - y, M) = (2\pi)^{-4}\int d^{4}p_{r}\, \exp[-ip^{0}(x^{0} - y^{0}) + i\mathbf{p_{r}}\cdot(\mathbf{x_{r}} - \mathbf{y_{r}})]/(p_{r}^{2} + M^{2} + i\varepsilon) \quad (3.214)$$

3.9 First Step towards a One Generation Standard Model

Thus we have found a set of four fermion species that corresponds to the known fermions of one fermion generation. In subsequent chapters we will derive the one generation model in detail. Then we will introduce three generations with mixing to complete the derivation of the form of the Standard Model. Then the only remaining major issue will be the values of the coupling constants and other numerical parameters.

The overall pattern that begins to emerge from the developments in this chapter divides particles and interactions into two categories (as seen in Nature):

Particles with real 4-Momenta	Complexons (Complex 3-Momenta)
Leptons	color quarks
SU(2)⊗U(1) Vector Bosons	Color SU(3) gluons
Higgs Particles	Possibly Higgs Particles

We will explore these issues in detail in the following chapters. But basically the leptons, SU(2)⊗U(1) Vector Bosons and a set of Higgs particles appear to be based on the Left-handed Extended Lorentz group. These particles have real energies and momenta although some are "normal" and some are tachyons.

The other category of particles, complexons, emerges from our study of L_C. These particles have real energies and complex 3-momenta. In perturbation theory the loop integrations of loops of these particles would consist of a 7-fold integration over energy and complex 3-momenta with corresponding 7-fold delta functions to enforce energy-momentum conservation. As pointed out earlier the complex 3-momenta of these types of fermions has an SU(3) symmetry that it is natural to generalize to local color SU(3). (The other category of fermions lacks this global SU(3) symmetry just as leptons lack color SU(3).) Thus we see the beginnings of the structure of the Standard Model in this chapter on spin ½ particles. The following chapters lead to a detailed derivation of the form of the Standard Model.

Appendix 3-A. Leptonic Tachyon Spinors

The general form of the solutions of the free tachyon Dirac equation eq. 3.18 can be written

$$\psi_T{}^r(x) = e^{-i\chi_r p\cdot x} w^r(p) \qquad (3\text{-A.1})$$

where $\chi_r = +1$ for r = 1, 2 and $\chi_r = -1$ for r = 3, 4. Denoting the spinors $w^r(p) = w^r(0)$ for a particle is at rest in a frame (E = m) we see they can take the form

$$w^r(0) = \begin{bmatrix} \delta_{1r} \\ \delta_{2r} \\ \delta_{3r} \\ \delta_{4r} \end{bmatrix} \qquad (3\text{-A.2})$$

where Kronecker deltas appear in the brackets. From eq. 3.15 we find

$$S_L(\Lambda_L(\omega, \mathbf{u}))w^r(0) = w_T{}^r(p) \qquad (3\text{-A.3})$$

Using eq. 3.11 for $S_L(\Lambda_L(\omega, \mathbf{u}))$ and

$$\mathbf{p} = m\mathbf{v}\gamma_s \qquad E = m\gamma_s \qquad (3\text{-A.4})$$

we see that eq. 3-A.3 implies the columns of the resulting $S_L(\Lambda_L(\omega, \mathbf{u}))$ matrix are

$$S_L(\Lambda_L(\omega, \mathbf{u})) = \begin{array}{cccc} \underline{w_T^{3}(p)} & \underline{w_T^{4}(p)} & \underline{w_T^{1}(p)} & \underline{w_T^{2}(p)} \end{array}$$
$$\begin{bmatrix} \cosh(\omega_L/2) & 0 & \sinh(\omega_L/2)p_z/p & \sinh(\omega_L/2)p_-/p \\ 0 & \cosh(\omega_L/2) & \sinh(\omega_L/2)p_+/p & -\sinh(\omega_L/2)p_z/p \\ \sinh(\omega_L/2)p_z/p & \sinh(\omega_L/2)p_-/p & \cosh(\omega_L/2) & 0 \\ \sinh(\omega_L/2)p_+/p & -\sinh(\omega_L/2)p_z/p & 0 & \cosh(\omega_L/2) \end{bmatrix}$$
(3-A.5)

based on the superluminal transformation of positive energy states to negative energy states (eqs. 3.15 and 3.16) with $p_\pm = p_x \pm ip_y$ and where $p = |\mathbf{p}|$. It is easy to verify

$$(i\not{p} - \chi_r m)w_T^{r}(p) = 0 \qquad (3\text{-A.6})$$

where $\chi_r = -1$ for r = 1, 2 and $\chi_r = +1$ for r = 3, 4.

The spinors that we defined in eq. 2.10 can be generalized in a manner similar to Dirac spinors. We will use a similar notation to the Dirac spinor notation:

$$\begin{aligned} u_T(p, s) &= w_T^{1}(p) \\ u_T(p, -s) &= w_T^{2}(p) \\ v_T(p, s) &= w_T^{3}(p) \\ v_T(p, -s) &= w_T^{4}(p) \end{aligned} \qquad (3\text{-A.7})$$

We define "double dagger" spinors:

$$\begin{aligned} u_T^{\ddagger}(p, s) &= u_T^{\dagger}(p, s)i\gamma\cdot\mathbf{p}/|\mathbf{p}| \\ u_T^{\ddagger}(p, -s) &= u_T^{\dagger}(p, -s)i\gamma\cdot\mathbf{p}/|\mathbf{p}| \end{aligned}$$

$$v_T^{\ddagger}(p, s) = v_T^{\dagger}(p, s)i\gamma\cdot\mathbf{p}/|\mathbf{p}| \qquad (3\text{-A.8})$$
$$v_T^{\ddagger}(p, -s) = v_T^{\dagger}(p, -s)i\gamma\cdot\mathbf{p}/|\mathbf{p}|$$

.

where † indicates hermitean conjugate, which appear in important spinor "completeness" sums:

$$\sum_{\pm s} u_{T\alpha}(p, s)u_{T\ \beta}^{\ \ddagger}(p, s) = (2m)^{-1}(i\not{p} - m)_{\alpha\beta} \qquad (3\text{-A.9})$$

$$\sum_{\pm s} v_{T\alpha}(p, s)v_{T\ \beta}^{\ \ddagger}(p, s) = (2m)^{-1}(i\not{p} + m)_{\alpha\beta} \qquad (3\text{-A.10})$$

or

$$\sum_{\pm s} u_{T\alpha}(p, s)u_{T\ \beta}^{\ \dagger}(p, s) = -i(2m)^{-1}[(i\not{p} - m)\gamma\cdot\mathbf{p}/|\mathbf{p}|]_{\alpha\beta} \qquad (3\text{-A.11})$$

$$\sum_{\pm s} v_{T\alpha}(p, s)v_{T\ \beta}^{\ \dagger}(p, s) = -i(2m)^{-1}[(i\not{p} + m)\gamma\cdot\mathbf{p}/|\mathbf{p}|]_{\alpha\beta} \qquad (3\text{-A.12})$$

Lastly we define light-front, left-handed tachyon spinors by

$$u_{TL}^{\ +}(p, s) = C^{-} R^{+} S_L(\Lambda_L(\omega, \mathbf{u}))w^1(0)$$
$$u_{TL}^{\ +}(p, -s) = C^{-} R^{+} S_L(\Lambda_L(\omega, \mathbf{u}))w^2(0) \qquad (3\text{-A.13})$$
$$v_{TL}^{\ +}(p, s) = C^{-} R^{+} S_L(\Lambda_L(\omega, \mathbf{u}))w^3(0)$$
$$v_{TL}^{\ +}(p, -s) = C^{-} R^{+} S_L(\Lambda_L(\omega, \mathbf{u}))w^4(0)$$

$$u_{TL}^{\ +\dagger}(p, s) = w^{1T}(0)\, S_L^{\dagger}(\Lambda_L(\omega, \mathbf{u}))\, R^{+}C^{-}$$
$$u_{TL}^{\ +\dagger}(p, -s) = w^{2T}(0)\, S_L^{\dagger}(\Lambda_L(\omega, \mathbf{u}))R^{+}C^{-} \qquad (3\text{-A.14})$$

$$v_{TL}^{+\dagger}(p, s) = w^{3T}(0)\, S_L^{\dagger}(\Lambda_L(\omega, \mathbf{u}))R^{+}C^{-}$$
$$v_{TL}^{+\dagger}(p, -s) = w^{4T}(0)\, S_L^{\dagger}(\Lambda_L(\omega, \mathbf{u}))R^{+}C^{-}$$

where the superscript "T" indicates the transpose and † indicates hermitean conjugate.

Appendix 3-B. Proof of the Reality of the Leptonic Tachyon Action

The tachyon lagrangian density and action are

$$\mathcal{L}_T = \psi_T{}^S(\gamma^\mu\partial/\partial x^\mu - m)\psi_T(x) \qquad (3.19)$$

$$I = \int d^4x\mathcal{L}_T \qquad (3.21)$$

where

$$\psi_T{}^S = \psi_T{}^\dagger\, i\gamma^0\gamma^5 \qquad (3.20)$$

The complex conjugate of the tachyon lagrangian density is

$$\mathcal{L}_T{}^* = -\psi_T{}^T\, i\gamma^0\gamma^5(\gamma^{\mu *}\partial/\partial x^\mu - m)\psi_T{}^*(x) \qquad (3\text{-B.1})$$

where the superscript ^{T} indicates the transpose. Eq. 3-B.1 can be expressed as a transpose:

$$\mathcal{L}_T{}^* = -i[\psi_T{}^\dagger(\gamma^{\mu\dagger}\overleftarrow{\partial}/\partial x^\mu - m)\gamma^5\gamma^0\psi_T(x)]^T \qquad (3\text{-B.2})$$

$$= -i[\psi_T{}^\dagger\gamma^5\gamma^0(-\gamma^\mu\overleftarrow{\partial}/\partial x^\mu - m)\psi_T(x)]^T \qquad (3\text{-B.3})$$

$$= [\psi_T{}^\dagger i\gamma^0\gamma^5(-\gamma^\mu\overleftarrow{\partial}/\partial x^\mu - m)\psi_T(x)]^T \qquad (3\text{-B.4})$$

$$= \psi_T{}^\dagger i\gamma^0\gamma^5(-\gamma^\mu\overleftarrow{\partial}/\partial x^\mu - m)\psi_T(x) \qquad (3\text{-B.5})$$

since eq. 3-B.4 is the transpose of a 1 by 1 matrix. Upon performing a partial integration in the action we find

$$I^* = \int d^4x[\psi_T{}^\dagger i\gamma^0\gamma^5(\gamma^\mu\partial/\partial x^\mu - m)\psi_T(x)] = I \qquad (3\text{-B.6})$$

REFERENCES

Akhiezer, N. I., Frink, A. H. (tr), 1962, *The Calculus of Variations* (Blaisdell Publishing, New York, 1962).

Bjorken, J. D., Drell, S. D., 1964, *Relativistic Quantum Mechanics* (McGraw-Hill, New York, 1965).

Bjorken, J. D., Drell, S. D., 1965, *Relativistic Quantum Fields* (McGraw-Hill, New York, 1965).

Blaha, S., 1998, *Cosmos and Consciousness* (Pingree-Hill Publishing, Auburn, NH, 1998).

_______, 2002, *A Finite Unified Quantum Field Theory of the Elementary Particle Standard Model and Quantum Gravity Based on New Quantum Dimensions™ & a New Paradigm in the Calculus of Variations* (Pingree-Hill Publishing, Auburn, NH, 2002).

_______, 2003, *A Finite Unified Quantum Field Theory of the Elementary Particle Standard Model and Quantum Gravity Based on New Quantum Dimensions™ and a New Paradigm in the Calculus of Variations* (Pingree-Hill Publishing, Auburn, NH, 2003).

_______, 2004, *Quantum Big Bang Cosmology: Complex Space-time General Relativity, Quantum Coordinates™ Dodecahedral Universe, Inflation, and New Spin 0, ½, 1 & 2 Tachyons & Imagyons* (Pingree-Hill Publishing, Auburn, NH, 2004).

______, *2005a, Quantum Theory of the Third Kind: A New Type of Divergence-free Quantum Field Theory Supporting a Unified Standard Model of Elementary Particles and Quantum Gravity based on a New Method in the Calculus of Variations* (Pingree-Hill Publishing, Auburn, NH, 2005).

______, 2005b, *The Metatheory of Physics Theories, and the Theory of Everything as a Quantum Computer Language* (Pingree-Hill Publishing, Auburn, NH, 2005).

______, 2005c, *The Equivalence of Elementary Particle Theories and Computer Languages: Quantum Computers, Turing Machines, Standard Model, Superstring Theory, and a Proof that Gödel's Theorem Implies Nature Must Be Quantum* (Pingree-Hill Publishing, Auburn, NH, 2005).

______, 2006a, *The Foundation of the Forces of Nature* (Pingree-Hill Publishing, Auburn, NH, 2006).

______, 2006b, *A Derivation of ElectroWeak Theory based on an Extension of Special Relativity; Black Hole Tachyons; & Tachyons of Any Spin.* (Pingree-Hill Publishing, Auburn, NH, 2006).

_______, 2007a, *Physics Beyond the Light Barrier: The Source of Parity Violation, Tachyons, and A Derivation of Standard Model Features* (Pingree-Hill Publishing, Auburn, NH, 2007).

______, 2007b, *The Origin of the Standard Model: The Genesis of Four Quark and Lepton Species, Parity Violation, the ElectroWeak Sector, Color SU(3), Three Visible Generations of Fermions, and One Generation of Dark Matter with Dark Energy* (Pingree-Hill Publishing, Auburn, NH, 2007).

______, *2008a, A Direct Derivation of the Form of the Standard Model From GL(16) (Pingree-Hill Publishing, Auburn, NH, 2008).*

______, 2008b, *A Complete Derivation of the Form of the Standard Model With a New Method to Generate Particle Masses Second Edition* (Pingree-Hill Publishing, Auburn, NH, 2008)

______, 2009, *The Algebra of Thought & Reality: The Mathematical Basis for Plato's Theory of Ideas, and Reality Extended to Include A Priori Observers and Space-Time Second Edition* (Pingree-Hill Publishing, Auburn, NH, 2009).

______, 2010a, *Operator Metaphysics: A New Metaphysics Based on a New Operator Logic and a New Quantum Operator Logic that Lead to a Mathematical Basis for Plato's Theory of Ideas and Reality* (Pingree-Hill Publishing, Auburn, NH, 2010).

______, 2010b, *The Standard Model's Form Derived from Operator Logic, Superluminal Transformations and GL(16)* (Pingree-Hill Publishing, Auburn, NH, 2010).

______, 2010c, *SuperCivilizations: Civilizations as Superorganisms* (McMann-Fisher Publishing, Auburn, NH, 2010).

______, 2011a, *21st Century Natural Philosophy Of Ultimate Physical Reality* (McMann-Fisher Publishing, Auburn, NH, 2011).

______, 2011b, *All the Universe! Faster Than Light Tachyon Quark Starships & Particle Accelerators with the LHC as a Prototype Starship Drive Scientific Edition* (Pingree-Hill Publishing, Auburn, NH, 2011).

______, 2011c, *From Asynchronous Logic to The Standard Model to Superflight to the Stars* (Blaha Research, Auburn, NH, 2011).

______, 2012a, *From Asynchronous Logic to The Standard Model to Superflight to the Stars volume 2: Superluminal CP and CPT, U(4) Complex General Relativity and The Standard Model, Complex Vierbein General Relativity, Kinetic Theory, Thermodynamics* (Blaha Research, Auburn, NH, 2012).

______, 2012b, *Standard Model Symmetries, And Four And Sixteen Dimension Complex Relativity; The Origin Of Higgs Mass Terms* (Blaha Reasearch, Auburn, NH, 2012).

______, 2013a, *Multi-Stage Space Guns, Micro-Pulse Nuclear Rockets, and Faster-Than-Light Quark-Gluon Ion Drive Starships* (Blaha Research, Auburn, NH, 2013).

______, 2013b, *The Bridge to Dark Matter; A New Sister Universe; Dark Energy; Inflatons; Quantum Big Bang; Superluminal Physics; An Extended Standard Model Based on Geometry* (Blaha Reasearch, Auburn, NH, 2013).

______, 2014a, *Universes and Megaverses: From a New Standard Model to a Physical Megaverse; The Big Bang; Our Sister Universe's Wormhole; Origin of the Cosmological Constant, Spatial Asymmetry of the Universe, and its Web of Galaxies; A Baryonic Field between Universes and Particles; Megaverse Extended Wheeler-DeWitt Equation* (Blaha Reasearch, Auburn, NH, 2014).

______, 2014b, *All the Megaverse! Starships Exploring the Endless Universes of the Cosmos Using the Baryonic Force* (Blaha Research, Auburn, NH, 2014).

______, 2014c, *All the Megaverse! II Between Megaverse Universes: Quantum Entanglement Explained by the Megaverse Coherent Baryonic Radiation Devices – PHASERs Neutron Star Megaverse Slingshot Dynamics Spiritual and UFO Events, and the Megaverse Microscopic Entry into the Megaverse* (Blaha Research, Auburn, NH, 2014).

______, 2015a, *PHYSICS IS LOGIC PAINTED ON THE VOID: Origin of Bare Masses and The Standard Model in Logic, U(4) Origin of the Generations, Normal and Dark Baryonic Forces, Dark Matter, Dark Energy, The Big Bang, Complex General Relativity, A Megaverse of Universe Particles* (Blaha Research, Auburn, NH, 2015).

______, 2015b, *PHYSICS IS LOGIC Part II: The Theory of Everything, The Megaverse Theory of Everything, U(4)⊗U(4) Grand Unified Theory (GUT), Inertial Mass = Gravitational Mass, Unified Extended Standard Model and a New Complex General Relativity with Higgs Particles, Generation Group Higgs Particles* (Blaha Research, Auburn, NH, 2015).

______, 2015c, *The Origin of Higgs ("God") Particles and the Higgs Mechanism: Physics is Logic III, Beyond Higgs – A Revamped Theory With a Local Arrow of Time, The Theory of Everything Enhanced, Why Inertial Frames are Special, Universes of the Mind* (Blaha Research, Auburn, NH, 2015).

______, 2015d, *The Origin of the Eight Coupling Constants of The Theory of Everything: U(8) Grand Unified Theory of Everything (GUTE), S^8 Coupling Constant Symmetry, Space-Time Dependent Coupling Constants, Big Bang Vacuum Coupling Constants, Physics is Logic IV* (Blaha Research, Auburn, NH, 2015).

______, 2016a, *New Types of Dark Matter, Big Bang Equipartition, and A New U(4) Symmetry in the Theory of Everything: Equipartition Principle for Fermions, Matter is 83.33% Dark,*

Penetrating the Veil of the Big Bang, Explicit QFT Quark Confinement and Charmonium, Physics is Logic V (Blaha Research, Auburn, NH, 2016).

______, 2016b, *The Periodic Table of the 192 Quarks and Leptons in The Theory of Everything: The U(4) Layer Group, Physics is Logic VI* (Blaha Research, Auburn, NH, 2016).

______, 2016c, *New Boson Quantum Field Theory, Dark Matter Dynamics, Dark Matter Fermion Layer Mixing, Genesis of Higgs Particles, New Layer Higgs Masses, Higgs Coupling Constants, Non-Abelian Higgs Gauge Fields, Physics is Logic VII* (Blaha Research, Auburn, NH, 2016).

______, 2016d, *Unification of the Strong Interactions and Gravitation: Quark Confinement Linked to Modified Short-Distance Gravity; Physics is Logic VIII* (Blaha Research, Auburn, NH, 2016).

______, 2016e, *MoND: Unification of the Strong Interactions and Gravitation II, Quark Confinement Linked to Large-Scale Gravity, Physics is Logic IX* (Blaha Research, Auburn, NH, 2016).

______, 2016f, *CQ Mechanics: A Unification of Quantum & Classical Mechanics, Quantum/Semi-Classical Entanglement, Quantum/Classical Path Integrals, Quantum/Classical Chaos* (Blaha Research, Auburn, NH, 2016).

______, 2016g, *GEMS: Unified Gravity, ElectroMagnetic and Strong Interactions: Manifest Quark Confinement, A Solution for the Proton Spin Puzzle, Modified Gravity on the Galactic Scale* (Pingree Hill Publishing, Auburn, NH, 2016).

______, 2016h, *Unification of the Seven Boson Interactions based on the Riemann-Christoffel Curvature Tensor* (Pingree Hill Publishing, Auburn, NH, 2016).

______, 2017a, *Unification of the Eleven Boson Interactions based on 'Rotations of Interactions'* (Pingree Hill Publishing, Auburn, NH, 2017).

______, 2017b, *The Origin of Fermions and Bosons, and Their Unification* (Pingree Hill Publishing, Auburn, NH, 2017).

______, 2017c, *Megaverse: The Universe of Universes* (Pingree Hill Publishing, Auburn, NH, 2017).

______, 2017d, *SuperSymmetry and the Unified SuperStandard Model* (Pingree Hill Publishing, Auburn, NH, 2017).

______, 2017e, *From Qubits to the Unified SuperStandard Model with Embedded SuperStrings: A Derivation* (Pingree Hill Publishing, Auburn, NH, 2017).

______, 2017f, *The Unified SuperStandard Model in Our Universe and the Megaverse: Quarks, …* , (Pingree Hill Publishing, Auburn, NH, 2017).

______, 2018a, *The Unified SuperStandard Model and the Megaverse SECOND EDITION A Deeper Theory based on a New Particle Functional Space that Explicates Quantum Entanglement Spookiness (Volume 1)* (Pingree Hill Publishing, Auburn, NH, 2018).

______, 2018b, *Cosmos Creation: The Unified SuperStandard Model, Volume 2, SECOND EDITION* (Pingree Hill Publishing, Auburn, NH, 2018).

______, 2018c, *God Theory (*Pingree Hill Publishing, Auburn, NH, 2018).

______, 2018d, *Immortal Eye: God Theory: Second Edition* (Pingree Hill Publishing, Auburn, NH, 2018).

______, 2018e, *Unification of God Theory and Unified SuperStandard Model THIRD EDITION* (Pingree Hill Publishing, Auburn, NH, 2018).

______, 2019a, *Calculation of: QED α = 1/137, and Other Coupling Constants of the Unified SuperStandard Theory* (Pingree Hill Publishing, Auburn, NH, 2019).

______, 2019b, *Coupling Constants of the Unified SuperStandard Theory SECOND EDITION* (Pingree Hill Publishing, Auburn, NH, 2019).

______, 2019c, *New Hybrid Quantum Big_Bang–Megaverse_Driven Universe with a Finite Big Bang and an Increasing Hubble Constant* (Pingree Hill Publishing, Auburn, NH, 2019).

______, 2019d, *The Universe, The Electron and The Vacuum* (Pingree Hill Publishing, Auburn, NH, 2019).

______, 2019e, *Quantum Big Bang – Quantum Vacuum Universes (Particles)* (Pingree Hill Publishing, Auburn, NH, 2019).

______, 2019f, *The Exact QED Calculation of the Fine Structure Constant Implies ALL 4D Universes have the Same Physics/Life Prospects* (Pingree Hill Publishing, Auburn, NH, 2019).

______, 2019g, *Unified SuperStandard Theory and the SuperUniverse Model: The Foundation of Science* (Pingree Hill Publishing, Auburn, NH, 2019).

______, 2020a, *Quaternion Unified SuperStandard Theory (The QUeST) and Megaverse Octonion SuperStandard Theory (MOST)* (Pingree Hill Publishing, Auburn, NH, 2020).

______, 2020b, *United Universes Quaternion Universe - Octonion Megaverse* (Pingree Hill Publishing, Auburn, NH, 2020).

______, 2020c, *Unified SuperStandard Theories for Quaternion Universes & The Octonion Megaverse* (Pingree Hill Publishing, Auburn, NH, 2020).

______, 2020d, *The Essence of Eternity: Quaternion & Octonion SuperStandard Theories* (Pingree Hill Publishing, Auburn, NH, 2020).

Eddington, A. S., 1952, *The Mathematical Theory of Relativity* (Cambridge University Press, Cambridge, U.K., 1952).

Fant, Karl M., 2005, *Logically Determined Design: Clockless System Design With NULL Convention Logic* (John Wiley and Sons, Hoboken, NJ, 2005).

Feinberg, G. and Shapiro, R., 1980, *Life Beyond Earth: The Intelligent Earthlings Guide to Life in the Universe* (William Morrow and Company, New York, 1980).

Gelfand, I. M., Fomin, S. V., Silverman, R. A. (tr), 2000, *Calculus of Variations* (Dover Publications, Mineola, NY, 2000).

Giaquinta, M., Modica, G., Souchek, J., 1998, *Cartesian Coordinates in the Calculus of Variations* Volumes I and II (Springer-Verlag, New York, 1998).

Giaquinta, M., Hildebrandt, S., 1996, *Calculus of Variations* Volumes I and II (Springer-Verlag, New York, 1996).

Gradshteyn, I. S. and Ryzhik, I. M., 1965, *Table of Integrals, Series, and Products* (Academic Press, New York, 1965).

Heitler, W., 1954, *The Quantum Theory of Radiation* (Claendon Press, Oxford, UK, 1954).

Huang, Kerson, 1992, *Quarks, Leptons & Gauge Fields 2nd Edition* (World Scientific Publishing Company, Singapore, 1992).

Jost, J., Li-Jost, X., 1998, *Calculus of Variations* (Cambridge University Press, New York, 1998).

Kaku, Michio, 1993, *Quantum Field Theory*, (Oxford University Press, New York, 1993).

Kirk, G. S. and Raven, J. E., 1962, *The Presocratic Philosophers* (Cambridge University Press, New York, 1962).

Landau, L. D. and Lifshitz, E. M., 1987, *Fluid Mechanics 2nd Edition*, (Pergamon Press, Elmsford, NY, 1987).

Misner, C. W., Thorne, K. S., and Wheeler, J. A., 1973, *Gravitation* (W. H. Freeman, New York, 1973).

Rescher, N., 1967, *The Philosophy of Leibniz* (Prentice-Hall, Englewood Cliffs, NJ, 1967).

Rieffel, Eleanor and Polak, Wolfgang, 2014, *Quantum Computing* (MIT Press, Cambridge, MA, 2014).

Riesz, Frigyes and Sz.-Nagy, Béla, 1990, *Functional Analysis* (Dover Publications, New York, 1990).

Sagan, H., 1993, *Introduction to the Calculus of Variations* (Dover Publications, Mineola, NY, 1993).

Sakurai, J. J., 1964, *Invariance Principles and Elementary Particles* (Princeton University Press, Princeton, NJ, 1964).

Sorokin, Pitirim, 1941, *Social and Cultural Dynamics* (Porter Sargent Publishers, Boston, MA, 1941).

Streater, R. F. and Wightman, A. S., 2000, *PCT, Spin, Statistics, and All That* (Princeton University Press, Princeton, NJ 2000).

Weinberg, S., 1972, *Gravitation and Cosmology* (John Wiley and Sons, New York, 1972).

Weinberg, S., 1995, *The Quantum Theory of Fields Volume I* (Cambridge University Press, New York, 1995).

Weinberg, S., 2000, *The Quantum Theory of Fields Volume III Supersymmetry* (Cambridge University Press, New York, 2000).

Weyl, H., 1950, *Space, Time, Matter* (Dover, New York, 1950).

Weyl, H., (Tr. S. Pollard et al), 1987, *The Continuum* (Dover Publications, New York, 1987).

INDEX

About the Author

Stephen Blaha is a well-known Physicist and Man of Letters with interests in Science, Society and civilization, the Arts, and Technology. He had an Alfred P. Sloan Foundation scholarship in college. He received his Ph.D. in Physics from Rockefeller University. He has served on the faculties of several major universities. He was also a Member of the Technical Staff at Bell Laboratories, a manager at the Boston Globe Newspaper, a Director at Wang Laboratories, and President of Blaha Software Inc. and of Janus Associates Inc. (NH).

Among other achievements he was a co-discoverer of the "r potential" for heavy quark binding developing the first (and still the only demonstrable) non-Aeolian gauge theory with an "r" potential; first suggested the existence of topological structures in superfluid He-3; first proposed Yang-Mills theories would appear in condensed matter phenomena with non-scalar order parameters; first developed a grammar-based formalism for quantum computers and applied it to elementary particle theories; first developed a new form of quantum field theory without divergences (thus solving a major 60 year old problem that enabled a unified theory of the Standard Model and Quantum Gravity without divergences to be developed); first developed a formulation of complex General Relativity based on analytic continuation from real space-time; first developed a generalized non-homogeneous Robertson-Walker metric that enabled a quantum theory of the Big Bang to be developed without singularities at $t = 0$; first generalized Cauchy's theorem and Gauss' theorem to complex, curved multi-dimensional spaces; received Honorable Mention in the Gravity Research Foundation Essay Competition in 1978; first developed a physically acceptable theory of faster-than-light particles; first derived a composition of extremums method in the Calculus of Variations; first quantitatively suggested that inflationary periods in the history of the universe were not needed; first proved Gödel's Theorem implies Nature must be quantum; provided a new alternative to the Higgs Mechanism, and Higgs particles, to generate masses; first showed how to resolve logical paradoxes including Gödel's Undecidability Theorem by developing Operator Logic and Quantum Operator Logic; first developed a quantitative harmonic oscillator-like model of the life cycle, and interactions, of civilizations; first showed how equations describing superorganisms also apply to civilizations. A recent book shows his theory applies successfully to the past 14 years of history and to *new* archaeological data on Andean and Mayan civilizations as well as Early Anatolian and Egyptian civilizations.

He first developed an axiomatic derivation of the form of The Standard Model from geometry – space-time properties – The Unified SuperStandard Model. It unifies all the known forces of Nature. It also has a Dark Matter sector that includes a Dark ElectroWeak sector with Dark doublets and Dark gauge interactions. It uses quantum coordinates to remove infinities that crop up in most interacting quantum field theories and additionally to remove the infinities that appear in the Big Bang and generate inflationary growth of the universe. It shows

gravity has a MOND-like form without sacrificing Newton's Laws. It relates the interactions of the MOND-like sector of gravity with the r-potential of Quark Confinement. The axioms of the theory lead to the question of their origin. We suggest in the preceding edition of this book it can be attributed to an entity with God-like properties. We explore these properties in "God Theory" and show they predict that the Cosmos exists forever although individual universes (or incarnations of our universe) "come and go." Several other important results emerge from God Theory such a functionally triune God. The Unified SuperStandard Theory has many other important parts described in the Current Edition of *The Unified SuperStandard Theory* and expanded in subsequent volumes.

Blaha has had a major impact on a succession of elementary particle theories: his Ph.D. thesis (1970), and papers, showed that quantum field theory calculations to all orders in ladder approximations could not give scaling deep inelastic electron-nucleon scattering. He later showed the eigenvalue equation for the fine structure constant α in Johnson-Baker-Willey QED had a zero at $\alpha = 1$ not 1/137 by solving the Schwinger-Dyson equations to all orders in an approximation that agreed with exact results to 4th order in α thus ending interest in this theory. In 1979 at Prof. Ken Johnson's (MIT) suggestion he calculated the proton-neutron mass difference in the MIT bag model and found the result had the wrong sign reducing interest in the bag model. These results all appear in Physical Review papers. In the 2000's he repeatedly pointed out the shortcomings of SuperString theory and showed that The Standard Model's form could be derived from space-time geometry by an extension of Lorentz transformations to faster than light transformations. This deeper space-time basis greatly increases the possibility that it is part of THE fundamental theory. Recently, Blaha showed that the Weak interactions differed significantly from the Strong, electromagnetic and gravitation interactions in important respects while these interactions had similar features, and suggested that ElectroWeak theory, which is essentially a glued union of the Weak interactions and Electromagnetism, possibly modulo unknown Higgs particle features, be replaced by a unified theory of the other interactions combined with a stand-alone Weak interaction theory. Blaha also showed that, if Charmonium calculations are taken seriously, the Strong interaction coupling constant is only a factor of five larger than the electromagnetic coupling constant, and thus Strong interaction perturbation theory would make sense and yield physically meaningful results.

In graduate school (1965-71) he wrote substantial papers in elementary particles and group theory: The Inelastic E- P Structure Functions in a Gluon Model. Phys. Lett. B40:501-502,1972; Deep-Inelastic E-P Structure Functions In A Ladder Model With Spin 1/2 Nucleons, Phys.Rev. D3:510-523,1971; Continuum Contributions To The Pion Radius, Phys. Rev. 178:2167-2169,1969; Character Analysis of U(N) and SU(N), J. Math. Phys. 10, 2156 (1969); and The Calculation of the Irreducible Characters of the Symmetric Group in Terms of the Compound Characters, (Published as Blaha's Lemma in D. E. Knuth's book: *The Art of Computer Programming Vols. 1 – 4*).

In the early 1980's Blaha was also a pioneer in the development of UNIX for financial, scientific and Internet applications: benchmarked UNIX versions showing that block size was critical for UNIX performance, developing financial modeling software, starting database benchmarking comparison studies, developing Internet-like UNIX networking (1982) and developing a hybrid shell programming technique (1982) that

was a precursor to the PERL programming language. He was also the manager of the AT&T ten-year future products development database. His work helped lead to commercial UNIX on computers such as Sun Micros, IBM AIX minis, and Apple computers.
In the 1980's he pioneered the development of PC Desktop Publishing on laser printers and was nominated for three "Awards for Technical Excellence" in 1987 by PC Magazine for PC software products that he designed and developed.
Recently he has developed a theory of Megaverses – actual universes of which our universe is one – with quantum particle-like properties based on the Wheeler-DeWitt equation of Quantum Gravity. He has developed a theory of a baryonic force, which had been conjectured many years ago, and estimated the strength of the force based on discrepancies in measurements of the gravitational constant G. This force, operative in D-dimensional space, can be used to escape from our universe in "uniships" which are the equivalent of the faster-than-light starships proposed in the author's earlier books. Thus travel to other universes, as well as to other stars is possible.
Blaha also considered the complexified Wheeler-DeWitt equation and showed that its limitation to real-valued coordinates and metrics generated a Cosmological Constant in the Einstein equations.
The author has also recently written a series of books on the serious problems of the United States and their solution as well as a book on the decline of Mankind that will follow from current social and genetic trends in Mankind.

In the past twenty years Dr. Blaha has written over 80 books on a wide range of topics. Some recent major works are: *From Asynchronous Logic to The Standard Model to Superflight to the Stars*, *All the Universe!*, *SuperCivilizations: Civilizations as Superorganisms, America's Future: an Islamic Surge, ISIS, al Qaeda, World Epidemics, Ukraine, Russia-China Pact, US Leadership Crisis, The Rises and Falls of Man – Destiny – 3000 AD: New Support for a Superorganism MACRO-THEORY of CIVILIZATIONS From CURRENT WORLD TRENDS and NEW Peruvian, Pre-Mayan, Mayan, Anatolian, and Early Egyptian Data, with a Projection to 3000 AD,* and *Mankind in Decline: Genetic Disasters, Human-Animal Hybrids, Overpopulation, Pollution, Global Warming, Food and Water Shortages, Desertification, Poverty, Rising Violence, Genocide, Epidemics, Wars, Leadership Failure.*

He has taught approximately 4,000 students in undergraduate, graduate, and postgraduate corporate education courses primarily in major universities, and large companies and government agencies.

Recently he developed a quantum theory, The Unified SuperStandard Theory (UST), which describes elementary particles in detail without the difficulties of conventional quantum field theory. He found that the internal symmetries of this theory could be exactly derived from a 32 dimension complex quaternion theory called QUeST. He further found that a 32 dimension complex octonion theory (MOST) describes the Megaverse. It can hold QUeST universes such as our own universe. It has an internal symmetry structure which is a superset of the QUeST internal symmetries.

www.ingramcontent.com/pod-product-compliance
Lightning Source LLC
Chambersburg PA
CBHW082007190326
41458CB00010B/3106
9781734583465